A Mathematical Primer on Groundwater Flow

An Introduction to the Mathematical and Physical Concepts of Saturated Flow in the Subsurface

John F. Hermance

Environmental Geophysics/Hydrology
Department of Geological Sciences
Brown University

Prentice Hall
Upper Saddle River, New Jersey 07458

Library of Congress Cataloging-in-Publication Data

Hermance, John F.
    A mathematical primer on groundwater flow : an introduction to the mathematical and physical concepts of saturated flow in the subsurface / John F. Hermance.
        p. cm.
    Includes bibliographical references and index.
    ISBN 0-13-896499-8
    1. Groundwater flow—Mathematical models. I. Title.
GB1197.7.H47 1998
551.49'01'5118—DC21

                                                  98-30038
                                                  CIP

Executive Editor: **ROBERT A. McCONNIN**
Editorial Assistant: **GRACE ANSPAKE**
Editorial/Production Supervision: **SUSAN FISHER**
Executive Managing Editor: **KATHLEEN SCHIAPARELLI**
Assistant Managing Editor: **LISA KINNE**
Manufacturing Manager: **TRUDY PISCIOTTI**
Director of Marketing: **LESLIE CAVALIERE**
Marketing Assistant: **RACHELE TRIANO**
Art Director: **JAYNE CONTE**
Cover Designer: **BRUCE KENSELAAR**

©1999 by John F. Hermance
Published by Prentice Hall, Inc.
Simon & Schuster / A Viacom Company
Upper Saddle River, New Jersey 07458

Cover Graphic: The figure on the cover represents the theoretical 1-D response in space and time of the hydraulic head of a confined aquifer (K = 10 m/day; S_s = 0.00025) for a transient, simple-harmonic forcing term at the tidal cycle (0.5 day). The hydraulic head has unit amplitude at the edge of the aquifer (x = 0), and decays exponentially with distance into the aquifer as discussed in Chapter 12 of the text.

Printed in the United States of America

10 9 8 7 6 5 4 3 2 1

ISBN 0-13-896499-8

Prentice-Hall International (UK) Limited, *London*
Prentice-Hall of Australia Pty. Limited, *Sydney*
Prentice-Hall Canada Inc., *Toronto*
Prentice-Hall Hispanoamericana, S.A., *Mexico*
Prentice-Hall of India Private Limited, *New Delhi*
Prentice-Hall of Japan, Inc., *Tokyo*
Simon & Schuster Asia Pte. Ltd., *Singapore*
Editora Prentice-Hall do Brasil, Ltda., *Rio de Janeiro*

Table of Contents

PREFACE

Purpose. This text bridges the gap between the essential relations of groundwater flow and many of the standard analytical techniques of applied mathematics. It is based on the premise that only by tracing the development of a mathematical relation from its underpinning physical laws through a specific sequence of clearly defined (usually simplifying) assumptions, can one appreciate the extent and limitations of how a particular formula can be applied to real-life situations. To some degree, this might seem to go counter to a current trend that would have students emphasize "solving practical problems" and for lecturers to "minimize the details of mathematical derivations" – in other words, for a course to down-play the formal mathematics in favor of more practical applications. In fact, I do not believe these two approaches need be at odds. Quite the contrary. This book is designed to resolve that conflict by presenting the mathematics in such a readable format that students can largely assimilate the material on their own, thus freeing instructors from having to formally develop much of the mathematical background through sacrificing precious lecture time. Of course, most hydrogeology courses these days are vectored toward numerical modeling in the later parts of the semester. In my experience, this text should provide the analytical "jolt" to get students up to speed by the sixth or seventh week of lectures, and – through a firm foundation in the analytical elements of gradient, divergence and Laplace's equation – allow an easy transition into finite differences in one, two and three dimensions.

A supplement to current texts. This book is *not* a comprehensive introduction to all aspects of hydrogeology; rather it is intended to *supplement* such standard texts as Fetter's "Applied Hydrogeology", Freeze and Cherry's "Groundwater", Walton's "Principles of Groundwater Engineering", Watson and Burnett's "Hydrology - An Environmental Approach", and Domenico and Schwartz's "Physical and Chemical Hydrology". In some cases, it may augment the material on groundwater flow in courses on more general aspects of physical hydrology based, for example, on Bras' "Hydrology"; Chow, Maidment and Mays' "Applied Hydrology"; and Dingman's "Physical Hydrology"; among others. I have used all of these texts in my courses and recommend each as well-written and readable. However, I have also come to deprecate the fact (while commiserating with the authors' concerns for economy) that these texts tend not to systematically derive the essential flow relations from first principles, but simply state them without proof. Since I, personally, feel that the derivations serve an end in themselves, I have found it useful to have material that can be handed off as supplementary reading to my students, or as background material for specific lectures, seminars or recitation groups. This is particularly true if I want my students to get into Wang and Anderson's "Introduction to Groundwater Modeling", or Anderson and Woessner's "Applied Groundwater Modeling".

Content. The present text attempts to connect the essential groundwater flow relations with the mathematical background that the quantitative student would have already been exposed to in his or her standard courses in physics, engineering and mathematics. I have found that students entering my introductory hydrology class have already come across concepts like boundary values, initial conditions, separation of variables, Fourier transforms, impulse responses, Green's functions and convolution in their various differential equations courses, but often do not connect the material to practical problems. This text purports to do so. The Theis equation (representing transient drawdown from a discharging well) is, after all, a simple restatement of the classical solution to transient diffusion in cylindrical coordinates. And the one dimensional transient flow solution for periodic sources looks a good deal like a solution to Maxwell's equations for dissipative media from electromagnetic theory. It is important for students to be aware of these parallels.

A "primer" not a treatise. This is *not* a treatise on mathematical hydrogeology. The broader views of mathematical analyses are already well done by such authors as Jacob Bear and Otto Strack. The present text is purposely short, treating only a minimum of the mathematical relations at the core of contemporary hydrogeology. It should be readable at the junior/senior undergraduate level and affordable as a supplement to one of the standard textbooks that would, and should, continue to serve as the basis for the overall course material, whether it be hydrogeology or broader aspects of physical hydrology. While some of the introductory, as well as some of the advanced, material is to be found in the more general texts mentioned above, this is done to make the present text relatively self-contained as a convenience to the reader. Most derivations are developed line by line, with little "left to the student". This is done purposely so the reader can actually *read* the text; perhaps not exactly like reading a novel, but then again, perhaps not having to put the book down every sentence or two in order to scribble out pages of mathematical manipulations on their own, or having to resort to a number of other textbooks to cross check their progress. It is no accident that, in a few places, I have taken several pages to explain what I have seen described in a mathematics text in only a few lines.

Intended audience. The ideal background for someone using this text would be mathematics through an introduction to partial differential equations, and a firm foundation in basic physics or engineering, preferably but not necessarily through electromagnetic theory. Academically, this book is primarily directed to undergraduate junior and senior physics, engineering and math concentrators, and to beginning graduate students in environmental science, geophysics and hydrology. The text should also appeal to professionals desiring to brush up on the mathematical side of groundwater flow on their own. However, even the reader *without* all the mathematical prerequisites, but who is still comfortable with the basics of differential and integral calculus, might well benefit from being exposed to where all that "good stuff" in math might lead, if indeed they decide, or can be prevailed upon, to take such courses at some point in the future. This text *does not* anticipate a background in geology, fluid mechanics or engineering hydraulics.

Acknowledgments

The concept for this text arose from discussions I had with Robert E. Hicks of Rampart Resources, a dear colleague and close friend, whose encouragement and example moved the material from course notes to completed text. The final product would be quite inadequate if not for the close reading of various drafts by my graduate students, Rabi Bohidar and Jeff Sullivan, as well as the undergraduate and graduate students in my various courses and seminars, notably Sang Won Lee, Kyler Eastman, Ben Hardt, and Scott Trull. Of particular note is the graciousness that William C. Walton, Herbert F. Wang and C. W. Fetter, Jr., extended to me during the nurturing stages of this material. I hope I can pass on their generous and open spirit to the up and coming hydrologists that approach *me* out of the blue.

The product you now see is much improved in scope and thought from the constructive comments of the reviewers of its draft versions: Herbert F. Wang, University of Wisconsin, Madison; James A. Tindall, U.S. Geological Survey, Denver; C.W. Fetter, Jr., University of Wisconsin, Oshkosh; Michael E. Campana, University of New Mexico, Albuquerque; Shemin Ge, University of Colorado, Boulder. Many thanks for your efforts.

All of us who have published through Prentice Hall, and have been exposed to the encouragement and supportive attitude of Bob McConnin, Executive Editor, are grateful for the opportunity to have worked with him, and wish him well in his retirement from behind the desk of the "Author's Editor". Finally, deepest thanks to Mary and Lou ...

Part I. Fundamental Relations of Groundwater Flow

Chapter 1. Hydrologic Nature of the Subsurface

Background

A Fundamental Conservation Condition

The conservation of matter is a fundamental concept in science and engineering. Hydrologists have adapted this notion to the paradigm of the "hydrologic cycle", which for the purposes of the present discussion can be stated, with reference to Figure 1.1, in the following integral form

$$\oiint_{\text{Surface}} \mathbf{q} \cdot d\mathbf{A} = \iiint_{\text{Vol}} [\text{All sources - All sinks}] \, dV \qquad (1.1)$$

With reference to Figure 1.1, the left hand side of the relation is an integral of the outward directed flux of water $\mathbf{q}$ through a closed mathematical surface, and the right hand side is the integral of all sources within the volume bounded by the closed surface. The flux $\mathbf{q}$ represents the volume of fluid crossing a cross-sectional area per unit time and $d\mathbf{A}$ is a local element of infinitesimal area represented as a vector directed along the local outward normal of the closed surface. In the SI system, $\mathbf{q}$ has units of $m^3/m^2/s$ or simply m/s, and $d\mathbf{A}$ has units of m^2.

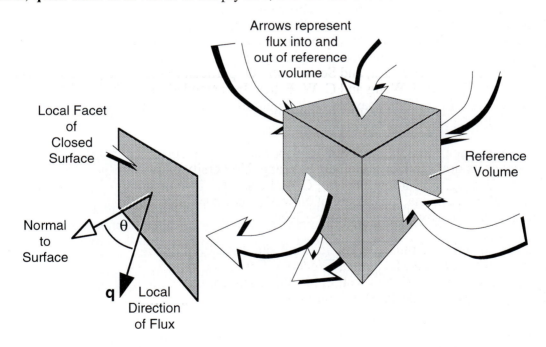

Figure 1.1 A conceptual view of the conservation integral in equation (1.1). The geometrical relation between $\mathbf{q}$, $d\mathbf{A}$ and θ, as discussed in the text, is shown for an arbitrary point on the left-facing surface.

It is emphasized that the closed surface is strictly a mathematical or hypothetical boundary in space. In general, it does not have any physical attribute for confining or enhancing flow, unless explicitly noted otherwise. To clarify the notation, $\mathbf{q}$ is the three dimensional vector flux given in terms of the volume of water passing through a unit area per unit time [$L^3 L^{-2} T^{-1}$ or $L\ T^{-1}$], and $d\mathbf{A}$ is the outward directed normal at a particular point on the closed surface. The term $\mathbf{q} \cdot d\mathbf{A}$ is the local vector dot or scalar product of $\mathbf{q}$ and $d\mathbf{A}$. This is the projection of $\mathbf{q}$ on to $d\mathbf{A}$, given by $|\mathbf{q}|\ |d\mathbf{A}|$ $\cos \theta$, where θ is the angle between the vector directions of $\mathbf{q}$ and $d\mathbf{A}$, the latter being directed along the outward normal to the surface.

Our intuitive notion of conservation is that if there are no sources or sinks within the volume, then whatever material flows *outward* through the closed surface ("positive" flux, since $\theta \leq 90^o$ and $\cos \theta \geq 0$) must be balanced by whatever material flows *inward* through the closed surface ("negative" flux, since $\theta \geq 90^o$ and $\cos \theta \leq 0$). Thus, in terms of the above integral, in the absence of sources and sinks within our reference volume, we have that

$$\oiint_{Surface} \mathbf{q} \cdot d\mathbf{A} = 0 \qquad (1.2)$$

which simply states that the net flux through the prescribed closed surface is zero — whatever flows in, flows out.

Local Hydrologic Cycle

Figure 1.2 shows a simple version of a local water cycle. There are basically two energy sources that "drive" the water cycle: solar radiation, which is responsible for evaporation, transpiration and atmospheric circulation; and the earth's gravity, which causes rain to fall (precipitation) and is responsible for infiltration (along with the intrinsic molecular attraction between water molecules and soil grains), surface runoff and groundwater flow.

Precipitation in the form of falling rain can be considered to be the "input function" into the groundwater system. During the initial stages of a moderate rain storm, moisture that reaches the ground is immediately absorbed by the soil; most of it adhering directly to soil grains through capillary action. At first, there is little preferential penetration of moisture to greater depth. As rainfall continues, there may be some puddling at the surface — so-called depression storage. And as the rain continues longer, some of the depression storage elements (puddles) may be breached, and overland flow develops along the surface. At some point following the initial "wetting" of the soil grains close to the surface, the volume of water collecting in the soil's interstitial voids becomes such that the body force on the fluid from gravitational attraction is stronger than the adhesion force that attracts and, in a sense, "binds" the bipolar water molecules to the atoms of the soil grains. As water is added to the system, additional layers of water molecules accumulate on the soil grains, but since these new water molecules are farther away from the grains, the attractive force will be diminished. When gravity dominates over the force of chemical adhesion, water molecules will tend to migrate to greater depth, infiltrating to the local water table where lateral groundwater flow generally occurs in a zone that is 100% saturated.

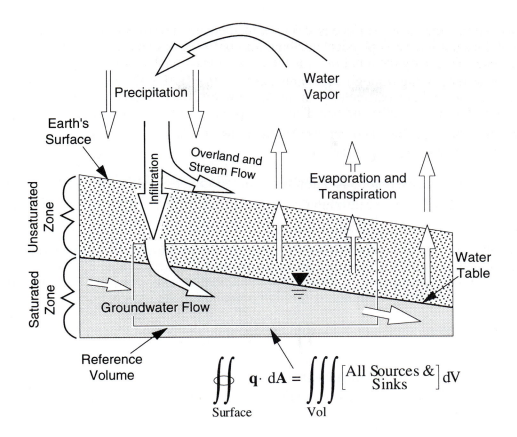

Figure 1.2 The local water cycle at the interface of the saturated zone. The conservation condition indicated by the integral at the bottom of the figure applies to the rectangular volume indicated. On the local scale, the hydrologic cycle is usually not closed; there is influx or outflux from the system.

In the case illustrated in Figure 1.2, water flows into the reference volume both as vertical infiltration from precipitation, and as lateral flow of groundwater from elsewhere in the system. If the flux of water out of the volume to the right side of the figure is just balanced by the input of infiltration through the top of the volume, and influx of groundwater through the left side of the volume, then we can assume that there are no sources or sinks in the volume, and relation (1.2) is obtained (i.e., the right hand side of (1.1) is zero). In the event that a non-negligible fraction of the infiltration entering the top of the reference volume adheres to the soil matrix within the volume or adds to a local increase in the elevation of the water table, there will be a net loss to the water budget passing outward through the enclosing surface. This additional interstitial water can be thought of as going into local "storage", or, alternatively, captured in a mathematical "sink". In this case, relation (1.1) would be a better representation of the situation, with the sink term on the right hand side being provided by the amount of water adhering to soil grains per unit volume per unit time or by the rate at which the water table rises in terms of the product of its average elevation increase per unit time, the horizontal area of the reference volume and the fractional porosity n of the medium ($0.0 \le n \le 1.0$). The negative sign for the sink accounts for the fact that there is a net loss of water passing *outward* through the surface of this reference volume.

There may be other pathways for water to be abstracted or removed from the reference volume. If, for example, plant or tree roots extend to this depth, there could be a vertical loss of water through the top surface by transpiration. Or there may be direct evaporation of water moisture from the interstitial pores.

In summary, the following fluid volumetric balance is obtained for a mathematical volume at the interface between the unsaturated and saturated zones in the subsurface. In the simplest terms:

$$[\text{ Flux(Out) - Flux(In) }] = \frac{\text{Flux produced (or lost)}}{\text{by sources in volume}} \tag{1.3}$$

To be more specific, we might have the following contributions:

$$\begin{array}{l} + \text{ Groundwater flow out} \\ - \text{ Groundwater flow in} \\ + \text{ Exfiltration} \\ - \text{ Infiltration} \end{array} = \frac{\text{Flux produced (or lost)}}{\text{by sources in volume}} \tag{1.4}$$

where the term "exfiltration" is used to denote all classes of unsaturated flux from the volume — this would include evaporation, transpiration through plant roots, and capillary action. In this text we will only consider saturated flow, although we will have occasion to discuss contributions to this flow through precipitation and losses through discharge to the surface.

THE SUBSURFACE ENVIRONMENT

Physical Foundations of Groundwater Flow

Darcy's law for pressure. Consider the experimental arrangement in Figure 1.3. This apparatus is known as a horizontal Darcy permeameter.

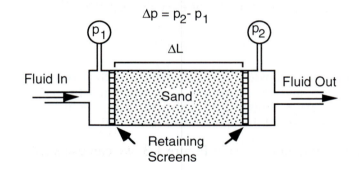

Figure 1.3 A horizontal Darcy permeameter. A device for measuring the hydrologic properties of a geologic material, in this case "sand".

For reasons that will become apparent later in our discussion, we will assume strictly horizontal flow, with water flowing in from the left at pressure p_1, and flowing out to the right at pressure p_2. The amount of water transferred per unit time Q (in m^3s^{-1}) is related to the "hydraulic conductivity" of the sand through the relation known as "Darcy's law for pressure":

$$Q = \frac{\text{Volume transferred}}{\text{Unit Time}} = - K_p \cdot \text{Area} \frac{\Delta p}{\Delta L} \tag{1.5}$$

where

    Q is the discharge [volume/unit time; m^3s^{-1}]

    Δp is the differential pressure [force/unit area; Newtons m^{-2} or Pascals]

ΔL is the length of the sample or the distance between the measuring points [m]

Area is the cross-sectional area of the sample [m²]

K_p is the hydraulic conductivity *for differential gauge pressure* [vol/unit time/unit cross-sectional area/unit pressure gradient; m³ s⁻¹ m⁻² (Pascals m⁻¹)⁻¹ or m² t⁻¹ Pascals⁻¹]

Note: The term $-\Delta p/\Delta L$ is the forcing term that "drives" fluid flow. It is analogous to the voltage gradient in electromagnetic theory, with Q being analogous to the total electric current I.

Darcy's law in terms of hydraulic head. We next consider an alternative formulation of Darcy's Law in terms of hydraulic head, rather than actual gauge pressure. Suppose instead of gauges, we employ "standpipes" to measure pressure, such as shown in Figure 1.4a.

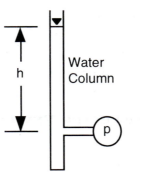

Figure 1.4a A "standpipe" to measure pressure.

In such a case, the pressure p (in Newton m⁻¹ or Pascals) is given by

$$p = \rho\, g\, h \tag{1.6}$$

where

ρ is the density of water [Kg m⁻³]

g is the gravitational force per unit mass [Newtons Kg⁻¹]

h is the height of the water column in meters (see Figure 1.4a)

Then instead of using pressure gauges, as in Figure 1.3, we can use standpipes, as in Figure 1.4b.

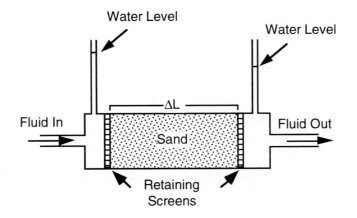

Figure 1.4b A Darcy permeameter using standpipes.

The difference in water level in these tubes is directly proportional to the difference in gauge pressure in Figure 1.3.

Pressure head and head loss. To be specific, we assume here and throughout the text that, unless stated otherwise, the term "pressure" will refer to the pressure measured with a gauge. A practical consideration of such instruments is that they record pressure with reference to atmospheric pressure which has a relative value of zero. By this convention, pressures less than one atmosphere are "negative". Following relation (1.6), we define "pressure head" as

$$h_p = p/\rho\, g \;=\; p/\gamma \;=\; \psi \qquad\qquad (1.7)$$

where

p is the gauge pressure (p = 0 at atmospheric pressure)

$\gamma = \rho\, g$ is the gravitational body force [units: Kg m^{-3} Newtons Kg^{-1}, or Newtons m^{-3}]

h_p and ψ are used interchangeably in the literature to denote the pressure head in meters at a point

Figure 1.5 illustrates various ways of measuring pressure head.

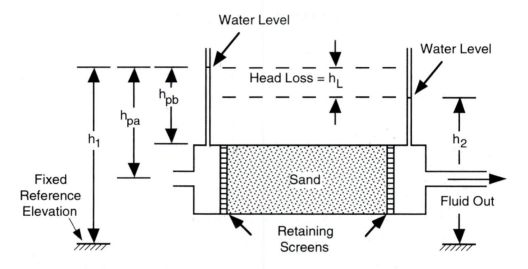

Figure 1.5 Various ways to measure "pressure head" using standpipes. "Head loss" is the difference in pressure head between two points.

With reference to Figure 1.5:

h_1 or h_2 is the water level relative to an arbitrary, but the same, fixed elevation. This parameter is usually referred to simply as the "hydraulic head".

h_{pa} is the water level relative to the elevation of the axis of the permeameter. It is the pressure head at a point on the axis of the permeameter, just outside the retaining screen.

h_{pb} is the water level relative to the elevation of the top of the permeameter. It is the pressure head at a point on the top of the permeameter, just outside the retaining screen.

"Head loss" h_L is the difference between the water level in two standpipes aligned in a particular direction, preferably in the principal direction of flow. This is usually given in terms of the difference in hydraulic head according to

$$h_L = \Delta h = h_1 - h_2 \qquad\qquad (1.8)$$

Darcy's law in h. Substituting the relation between gauge pressure p and hydraulic head h given by (1.6) into the differential pressure Δp, we obtain

$$\Delta p = p_2 - p_1 = \rho\, g\, h_2 - \rho\, g\, h_1 = \gamma \Delta h \qquad (1.9)$$

Substituting (1.9) for Δp in Darcy's law (1.5) leads to the following expression for the total discharge Q

$$Q = \frac{\text{Volume transferred}}{\text{Unit Time}} = -K_p \cdot \gamma \cdot \text{Area} \frac{\Delta h}{\Delta L} \qquad (1.10)$$

If we define $K_h = K_p \gamma$ as the *hydraulic conductivity for head differences*, Darcy's law can be expressed in terms of the hydraulic head

$$Q = -K_h \cdot \text{Area} \frac{\Delta h}{\Delta L} \qquad (1.11)$$

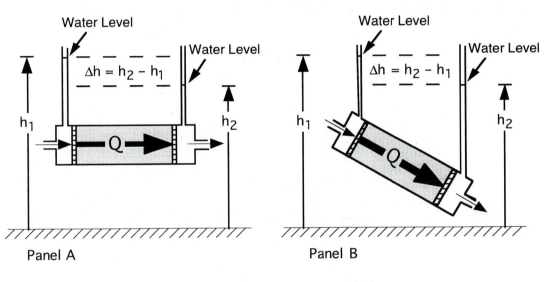

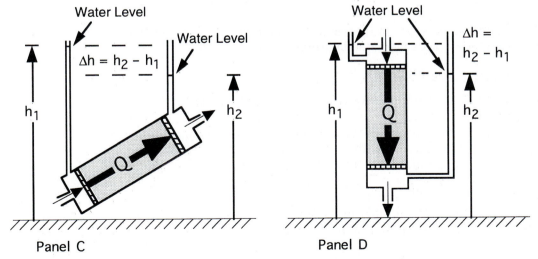

Figure 1.6 Four examples for which the total flow Q is equal, as long as the head loss is the same and gravity is uniform.

Hubbert (1940) argued that expressing Darcy's law in terms of differences in hydraulic head (e.g. (1.11)), rather than differences in gauge pressure or local pressure head, is applicable, not only to horizontal flow as discussed above, but to all flow directions in general. This is discussed in detail in a later section, but for now we will use a simple empirical result that Hubbert first posed, and is illustrated in Figure 1.6.

For the present purpose, we neglect changes of gravity over the scale of our experiment. If we employ the same or an identical permeameter containing the same or identical material, then we are assured that the cross-sectional area, the length ΔL and the hydraulic conductivity K_h are all constant parameters of the experiment illustrated in the figure. The two variables remaining in (1.11) are the total discharge Q and the difference in hydraulic head.

Now suppose we position the permeameter in various directions, assuring that the total discharge is the same for each configuration. Panel A is a reproduction of the horizontal flow experiment discussed above; Panels B and C are for the permeameter inclined at +/-30° to the horizontal; Panel D is for strictly vertical flow. In all cases, if Q is the same, then one would observe the same difference in hydraulic head $\Delta h = h_2 - h_1$. Δh would be negative, and flow would proceed from regions of high hydraulic head to regions of low hydraulic head. This result carries over to distributed flow in natural systems, and provides a fundamental cornerstone for our later analysis.

Specific discharge. Let the rate of fluid transport per unit cross-sectional area of the permeameter be given by q, defined as

$$q = Q/Area \qquad (1.12)$$

where Q is the volume flow in terms of the total volume of fluid discharged from the permeameter per unit time — such as gallons per minute, or cubic meters per second. "Area" refers to the cross-sectional area of the device measured perpendicular to the flow direction.

The units of q are volume/unit area/unit time, which is equivalent to length/unit time (m s^{-1} in the SI system), or velocity. In the literature, q is known by various names:

Darcy velocity
Darcy flux
Specific discharge
Filtration velocity

In this text, we will favor the terms "specific discharge" or "flux".

Assume a coordinate system oriented such that the x axis is along the axis of the permeameter. In terms of the specific discharge q, and differential pressures along the x axis, Darcy's law for pressure becomes:

$$q = - K_p \, \Delta p/\Delta x \qquad (1.13)$$

where Δx is an increment of length or distance, and K_p is the hydraulic conductivity in appropriate units when gauge pressure is used.

Employing the hydraulic head h, Darcy's law for hydraulic head becomes:

$$q = - K_h \, \Delta h/\Delta x \qquad (1.14)$$

where K_h is the hydraulic conductivity in appropriate units when the pressure head is used. In the limit of small Δx, we have the differential expression:

$$q = -K_p \, dp/dx \qquad (1.15)$$

Or in terms of the hydraulic head h

$$q = -K_h \, dh/dx \qquad (1.16)$$

Note that K_p and K_h, while both are considered to be hydraulic conductivities, have different magnitudes and units. In short, since q is the same in both (1.15) and (1.16), we can equate the right hand side of both equations, and use (1.6) to solve for

$$K_p = \frac{K_h}{\rho g} = \frac{K_h}{\gamma} \qquad (1.17)$$

K_h has the units of [m s^{-1}], whereas K_p has the units of [m s^{-1} (newton m^{-3})$^{-1}$]; [m^4 s^{-1} N^{-1}]; or [m^3 s Kg^{-1}].

•　　•　　•

Aside: **Analogy with electromagnetic theory**

Relation (1.16) is quite analogous to Ohm's law from electromagnetic theory:

$$J = -\sigma \, dV/dx \qquad (a)$$

or

$$J = \sigma E \qquad (b)$$

where

　　J is the electric current density (amperes/m^2)
　　σ is the electrical conductivity (siemens/m; the inverse of resistivity in ohm-m)
　　E is the electric field intensity (volts/m), given in terms of the voltage gradient by

$$E = -dV/dx \qquad (c)$$

End of aside.

•　　•　　•

Hydraulic Conductivity in Terms of the Intrinsic Physical Properties of the Medium and the Fluid

Case of an ideal material. Consider controlled experiments (following Freeze and Cherry, 1979, p. 26) using glass beads of diameter d to represent the medium in a permeameter of the type shown in Figure 1.4b, and fluids of various weight-densities ρg and dynamic viscosities μ. The latter parameter, in the SI system, has units of poise, where one poise is equivalent to one newton s m^{-2}

For a given experiment, the total discharge Q [in units of volume/unit time or m^3 s^{-1}] is related to the principal parameters of the procedure in the following way:

$Q \propto \Delta h$, the differential hydraulic head.

$Q \propto A$, the cross-sectional area.

$Q \propto 1/\Delta L$, where ΔL is the distance between the points at which the pressure is determined.

$Q \propto d^2$, the mean grain diameter-squared.

$Q \propto \rho g$, the gravitational body force on a unit volume of fluid.

$Q \propto 1/\mu$, where μ is the viscosity of the fluid. It is this latter term that reflects the first order dependence of fluid flow on the temperature — viscosity depends exponentially on inverse T.

Recalling that the specific discharge is given by

$$q = Q/\text{Area} \tag{1.12}$$

The above inter-relations can be combined into the form

$$q = - [Cd^2 \rho g/\mu] \, \Delta h/\Delta L \tag{1.18}$$

where C is a dimensionless numerical constant depending on the properties of the host matrix. One can show that C is dimensionless by rearranging (1.18) to the form

$$C = \frac{q \, \mu}{\rho g \, d^2} \frac{\Delta L}{\Delta h}$$

which has the dimensions

$$[C] = \left[\frac{q \, \mu}{\rho g \, d^2} \frac{\Delta L}{\Delta h} \right] = \left[\frac{\left(L \, T^{-1} \right)\left(M \, L^{-1} \, T^{-1} \right)}{\left(M \, L^{-2} \, T^{-2} \right)\left(L^2 \right)} \frac{L}{L} = 1 \right]$$

Moreover, in comparing (1.18) with (1.16), it is clear that the hydraulic conductivity for head differences is given by

$$K_h = [Cd^2 \rho g/\mu] \tag{1.19}$$

In this expression, the "body force" ρg on a volume of liquid, and the dynamic viscosity μ are properties of the fluid alone, whereas the product Cd^2 is a property of the medium alone. The latter term we define as the *permeability* of the medium, and denote it by

$$k = Cd^2 \tag{1.20}$$

All of the effects of grain shape and size, their relative positions, interconnectedness etc., are embodied in measurements of the permeability. In the SI system, k has units of m^2. Another unit of permeability, often used in the groundwater literature, is the "darcy", which is not an SI unit. The conversion factor to the SI system is

$$1 \text{ darcy} = 0.987 \times 10^{-12} \text{ m}^2 \tag{1.21}$$

Case of real, non-ideal materials. For the non-ideal case of materials consisting of a mixture of actual grains having different sizes, one can use (1.20) as a conceptual template on which to base the following empirical relation:

$$k = C_{char} \, d_{char}{}^{m} \qquad\qquad (1.22)$$

where

C_{char} = Characteristic shape factor [dimensionless],

d_{char} = Characteristic grain size [m],

m = Empirically determined exponent (perhaps non-integer).

SATURATED AND UNSATURATED CONDITIONS IN THE SUBSURFACE

Fundamental Volumetric Parameters

Consider the following partitioning of volume space:

V_t is the total volume of the sample.
V_v is the volume of the pore space or voids.
V_s is the volume of the solid material.
V_w is the volume of the water in the voids.
V_d is the volume of the water that drains from a saturated sample under the influence of gravity.
V_r is the volume of the water retained by a previously saturated sample following gravity drainage.

Porosity, Void Ratio, and Moisture Content

The *porosity* n of a sample is defined as the ratio of the volume of its pore space or voids V_v to its total volume V_t.

$$n = V_v \, / \, V_t \qquad\qquad (1.23)$$

Porosity for materials having uniform sized grains. The geometric packing structure of grains plays a significant role in a materials porosity. A simple cluster of spheres having identical radii can be packed in six possible ways resulting in porosities of approximately 26%, 30%, 40% and 48% (de Marsily, 1986).

Note: For such uniform grain-size distributions, the resulting porosity depends only on the packing configuration and is independent of the actual grain size. However, as discussed in an earlier section, the hydraulic conductivity, or permeability, will vary in direct proportion to the square of the grain size. Thus materials with high porosity are not always highly permeable.

Case of ideal non-uniform sized grains. If different sized spheres are mixed together, smaller spheres will tend to "fill in" the voids between larger spheres, reducing the overall porosity. Thus poorly sorted materials having a broad range of grain sizes will have low porosities.

Void ratio. The *void ratio* e is defined as the ratio of the pore volume or void volume V_V to the volume of the material comprising the solid matrix V_S.

$$e = V_V / V_S. \tag{1.24}$$

The relation between porosity and void ratio is given by

$$e = n / (1-n), \tag{1.25}$$

or

$$n = e / (1+e) \tag{1.26}$$

Saturation. A material is 100% saturated when the water volume V_w is equal to the void volume V_v. The volumetric saturation (in fractional percent) of a material is given by

$$s = V_w / V_v \tag{1.27}$$

Moisture content. The volumetric moisture content is given by

$$\theta = V_w / V_t \tag{1.28}$$

For saturated conditions, $\theta = n$.
For unsaturated conditions, $\theta < n$.

Yield and Retention of Groundwater

Specific yield. Under gravity drainage, the volume of water released under gravity drainage (V_d) from a volume of initially saturated material, divided by the total volume of the material (V_t),

$$S_y = V_d / V_t \tag{1.29}$$

Specific retention. Under gravity drainage, the volume of water retained (V_r) by a volume of an originally saturated material divided by the original volume of the material,

$$S_r = V_r / V_t \tag{1.30}$$

Clearly, the sum of the specific yield and the specific retention equals the porosity:

$$n = S_y + S_r \tag{1.31}$$

$$= (V_d / V_t) + (V_r / V_t) = (V_d + V_r) / V_t \tag{1.32}$$

Whereas, for a theoretically ideal sediment having a uniform grain size, the porosity depends only on the packing configuration and is independent of the grain size, such is not the case for the specific yield or specific retention.

Water is retained by a sediment due to the molecular attraction between the atoms comprising the solid material of the grains and the bipolar water molecules. Since this is largely a surface effect, materials higher surface area to pore volume ratios will generally have stronger retentivities. Thus, for equal porosities, materials having smaller mean grain sizes will tend to have larger specific retentions (or smaller specific yields).

The Saturated Zone

Under normal conditions, material immediately beneath the surface is unsaturated. However, at some depth, even for the driest terrains, one will encounter water molecules in the interstitial pores or cracks of the host material, and in most cases this water will be free to migrate under normal hydraulic stresses. Here we define some of the terms and concepts related to the saturated zone with reference to Figure 1.7.

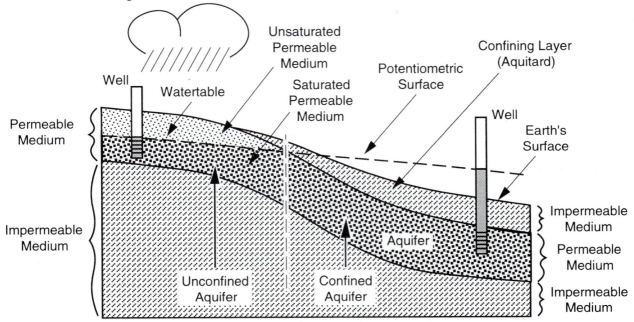

Figure 1.7 Schematic illustrating the transition from an unconfined aquifer on the left to a confined aquifer on the right. The "recharge area" is on the left and the water level is shown for two representative wells, with the potentiometric surface (discussed in a later section) shown as a dashed line. An aquitard, aquifuge or aquiclude might serve as the "confining" layer above the confined aquifer.

Water table. The water table is a free surface on which the fluid pressure p is precisely atmospheric (or $p = 0$ in gauge pressure). The water table is indicated by the shallowest level at which water stands in an open well which barely penetrates the water bearing deposit. Saturated conditions are obtained at and below the water table. But in some cases, capillary forces are responsible for partially-saturated conditions *above* the operationally defined water table.

Aquifer, aquiclude, aquifuge, and aquitard. These terms are strictly relative.

Aquifer: Operationally defined as any geologic unit or feature that can store or supply water in usable quantities. "Usable" quantities is the relative term.

Aquiclude: The low permeability upper or lower boundary of an aquifer.

Aquifuge : A material that is absolutely impermeable.

Aquitard : A material that has low permeability, but can transmit water at low flow rates.

One should be aware that the above definitions are somewhat arbitrary. A silt layer between gravel layers would normally be considered an "aquitard" or even an "aquifuge" for a public supply aquifer. On the other hand, a silt layer between clay layers could be an "aquifer" for a low production domestic well.

Modern usage tends to use only the terms *aquifer* and *aquitard* (Freeze and Cherry, 1979).

Unconfined versus confined aquifers. An *unconfined* aquifer (also known as a *water table* aquifer) is one in free communication with the atmosphere.

A *confined* aquifer (also known as an *artesian* aquifer) is one that is bounded above and below by impermeable layers (termed *confining layers*), and whose fluid has a hydraulic head exceeding the elevation of the top of the aquifer. These confining layers may be totally impermeable aquifuges, or low permeability aquitards. The latter case is also known as a *leaky confining layer*. In operational terms, fluid in a confined aquifer is under sufficient pressure to force it to rise in a well bore above the top of the aquifer.

Throughout most of this text, we will be even more specific in the qualities of the confined aquifers that we will consider. Unless indicated otherwise, we will assume for mathematical simplicity that the aquifer is horizontal in the x-y plane, having a uniform thickness b, and a uniform hydraulic conductivity K. We will also generally assume that any recharge or discharge at a particular point $P(x,y)$ takes place instantaneously and uniformly over the entire thickness of the aquifer.

Chapter 2. Darcy's Law and Three Dimensional Flow

DETAILED CONSIDERATIONS OF DARCY'S LAW

Darcy's Law in Terms of Pressure Head

In the previous chapter, we considered Darcy's law for gauge pressure, pressure head, and hydraulic head, respectively. We stated that, of these three forms, Darcy's law in terms of the hydraulic head was the most generally useful. We now explore the reasons for this.

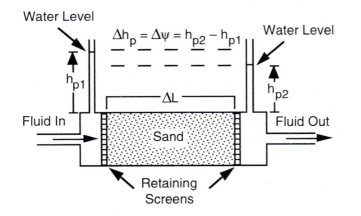

Figure 2.1 A Darcy experiment for horizontal flow using measurements of the local pressure head $\psi = h_p$ in standpipes.

We begin by considering Darcy's Law for the total discharge Q in terms of the differential pressure head Δh_p, in the form

$$Q = - K \cdot \text{Area} \cdot \Delta h_p / \Delta L \qquad (2.1)$$

where

Q is the total discharge [having dimensions of unit volume/unit time $= L^3 T^{-1}$],

K is the hydraulic conductivity, which throughout the remainder of this text will be understood to represent the hydraulic conductivity for differential heads, so that $K \equiv K_h$ [having dimensions of $L T^{-1}$],

Area is the cross-sectional area of the vessel $[L^2]$,

Δh_p is the differential pressure head, locally determined using standpipes [L],

ΔL is the length of the sample [L].

Such an experimental procedure works fine for horizontal flow, but, as we see in the following discussion, fails without proper modification for more general cases.

Hubbert's Conundrum

Hubbert (1940) has gone to some detail in discussing the difficulties that can arise if one is not careful in defining what they assume to be the fundamental driving forces for water flow — the pressure differential terms Δp or Δh. (For excellent accessible discussions one should see Freeze and Cherry, 1979; or Fetter, 1994). The problem is relatively unambiguous for strictly horizontal flow, but if flow deviates from the horizontal, one needs to exercise some care in their definition of pressure and/or hydraulic head differences.

Hubbert (1940) underscored this point using the following argument. Suppose one were to define the "driving force" as the difference between pressure heads measured with standpipes at the in-flow point 1 and the out-flow point 2, respectively. While, as we saw for the case illustrated in Figure 2.1, there is no problem with this procedure for horizontal flow, if one tilts their vessel to the case shown in Figure 2.2, they have the rather puzzling result that

$$h_{p1} < h_{p2} \qquad (2.2)$$

or that the pressure head at point 1 *is less than* the pressure head at point 2. According to this, water is flowing from a region of *low* pressure head to a region of *high* pressure head; quite contrary to our intuition. A little thought will indicate that this is not a problem unique to standpipes; pressure gauges would lead to the same result. This is what we call the "Hubbert Conundrum": a thought-puzzle posed by King Hubbert in 1940 to underscore the importance of properly understanding the underlying physical and mathematical principles of saturated flow.

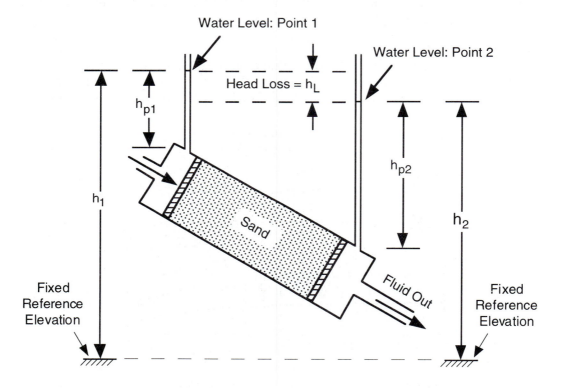

Figure 2.2 The basis for Hubbert's Conundrum: A Darcy experiment for non-horizontal flow using measurements of the local pressure head $\psi = h_p$ in standpipes leads to inconsistent and erroneous results (see text).

On the other hand, Figure 2.2, shows that the head loss h_L — the difference in elevation of the water levels in the two standpipes — is such that the water level at point 2 would be at a lower elevation than at point 1. Moreover, if we employed the "hydraulic heads" measured to a fixed reference elevation as shown in this figure as h_1 and h_2, respectively, we should have

$$h_1 > h_2 \qquad (2.3)$$

so that, according to these last two observations, water appears to flow from a region of high "pressure" to a region of low "pressure", which accords with our intuition, but is opposite to what was inferred from the pressure heads discussed above. How do we reconcile these results?

PRESSURE, ELEVATION HEAD, HYDRAULIC HEAD, PRESSURE HEAD, TOTAL MECHANICAL ENERGY, AND FLUID POTENTIAL

As Hubbert (1940) argued, in order to unambiguously relate Darcy flow to physically measurable parameters, one needs to rigorously understand the underlying physical processes involved. We first define several relevant terms in the context of field-measurable parameters.

Pressure: A scalar function of position representing force (Newtons) per unit area (m^2). Pressures are often implicitly given in relative terms, e.g. relative to standard atmospheric pressures p_o, or relative to a value measured elsewhere in the system.

In Figure 2.3, we set the stage for defining terms that will be used throughout the remainder of the text. These are (referring to the figure):

h is the "hydraulic head";
ψ or h_p will be used interchangeably, depending on context, for the "pressure head";
z is the local elevation of a point P in the subsurface relative to a global reference system. (Often this reference surface is sea level, but it can be defined in some cases as a convenient local "datum".)

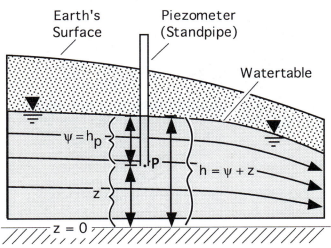

Figure 2.3 A standpipe is inserted into the subsurface to sense conditions at a point P in an aquifer.

The watertable is indicated by the inverted triangles. A reference plane is assumed at $z = 0$; the elevation of the point P above the reference surface is z; and the local pressure head is the height or elevation of the water surface in the standpipe above P, or $\psi = h_p$.

Elevation Head: The elevation or height z of a point P above some reference surface, which with no loss of generality we set equal to zero (i.e., $z = 0$); it could be set to any arbitrary constant value.

Hydraulic Head: Assume that an open standpipe (also referred to by various authors as a "manometer", "piezometer", or "potentiometer") has its lower end at a point in a water-saturated, permeable material (see also Figure 1.7). The hydraulic head h at that point is defined as the height above a standard reference surface to which a column of water will rise in the standpipe.

Pressure Head: A locally, referenced parameter evaluated at a point P in a water-saturated, permeable material. It is defined as the height of the water column in a standpipe relative to the vertical position z of P:

$$\psi = h_p = h - z \tag{2.4}$$

Total Mechanical Energy: The total mechanical energy E_{tot} per unit volume of mass ρ, under pressure p, at a point P, moving with a velocity v, in a gravitational field g (force per unit mass) can be represented as the sum of three sub-components of the energy of the system:

$E_g = \rho g z$, gravitational energy;

$E_v = \rho v^2/2$, kinetic energy (usually negligible);

$E_p = (p - p_o)$, energy expended or work done on a unit volume in raising the fluid pressure from p_o to p *at the point* P.

Thus, we have

$$E_{tot} = E_g + E_v + E_p \tag{2.5}$$

Usually v is small for fluid migration in aquifers, so the kinetic energy term E_v can be neglected.

Fluid Potential: The fluid potential (or force potential) is defined as the total mechanical energy per unit mass,

$$W = E_{tot} / \rho \tag{2.6}$$

or

$$W = gz + (p - p_o)/\rho \tag{2.7}$$

The measurement of p relative to atmospheric pressure is so commonly done that the term p_o is simply assumed and suppressed, thus we can write the fluid potential as

$$W = gz + p/\rho \tag{2.8}$$

To reiterate, this is the total mechanical energy per unit mass at the point P (at a height z above the reference plane).

Fluid Potential in Terms of Hydraulic Head

The above expression for the fluid potential W can be simplified using the following definition for the relative pressure in terms of the local pressure head ψ:

$$p = \rho g \psi \tag{2.9}$$

Substituting this into the previous expression, leads to

$$W = gz + (\rho g \psi)/\rho \tag{2.10}$$

and employing

$$\psi = h - z \qquad\qquad (2.11)$$

we obtain

$$W = gz + [\rho g(h - z)]/\rho \qquad\qquad (2.12)$$

which reduces to

$$W = gh \qquad\qquad (2.13)$$

This states that the fluid potential of any point P is independent of the vertical position of the point, but depends only on the product of the hydraulic head and the force of gravity.

Darcy's Law in Terms of Fluid Potential

The fluid or force potential W (or the hydraulic head h) is a fundamentally useful parameter to describe the motion of fluid in the subsurface. To summarize the extensive discussion of Hubbert (1940) on the matter, the most unambiguous form of Darcy's law is in terms of the fluid potential:

For a homogeneous fluid of constant density, flux is in the direction from higher values of the fluid potential to lower values of the fluid potential, that is to say, in the direction of the negative gradient of the fluid potential.

Darcy's law for the total discharge from the apparatus in Figure 2.2 can accordingly be written in the form

$$Q = - K_W \cdot Area \frac{\Delta W}{\Delta L} \qquad\qquad (2.14)$$

where K_w is the hydraulic conductivity for differential fluid potential.

In terms of the specific discharge, $q = Q / Area$, (2.14) leads to

$$q = - K_W \frac{\Delta W}{\Delta L} \qquad\qquad (2.15)$$

In considering the fluid potential

$$W = gh \qquad\qquad (2.13)$$

it is usual to assume that g (the gravitational force per unit mass) is constant. (We already have assumed that the density of the fluid ρ is constant, or we should return to posing our relations in terms of the total energy $E_{tot} = \rho W$.) Assuming g is constant is certainly reasonable for laboratory, local and most regional scale problems; but one should keep this assumption in mind as they approach problems of planetary proportions. Thus, providing g is constant, we define

$$K = K_W \cdot g \qquad\qquad (2.16)$$

(where, recall, $K = K_h$) so that expression 2.15 leads to the following expression for Darcy's law for specific discharge:

$$q = \frac{Q}{\text{Area}} = -K\frac{\Delta h}{\Delta L} \tag{2.17}$$

Darcy's law, in the form of (2.17), is a rigorous statement based on energy considerations. It underscores the importance of defining the "driving force" of fluid flow as being due to differences in the hydraulic head; not to other parameters such as the local pressure head that lead to "Hubbert's Conundrum". When considering the flow of homogeneous fluids of constant density, one is best off using the hydraulic head as the driving parameter. If, for some reason, one has to resort to other parameters, such as the local gauge pressure p, or the local pressure head $\psi = h_p$, (such as is often done in mechanical piping and/or plumbing) one needs to refer their observations to pre-existing, "no-flow" static values. While practical for Darcy permeameters and other controlled situations, this is usually not possible under natural conditions. Therefore, most hydrological analyses are cast in terms of the hydraulic head h.

Flow in Distributed Media

By distributed media, we mean those situations, usually natural, where material properties and flow conditions are not strictly controlled in a contained vessel under laboratory conditions. In many cases, the hydraulic conductivity is a scalar function of three spatial dimensions, such that

$$K = K(x,y,z) \tag{2.18}$$

In the most general case, fluid may flow in three dimensions, so that the specific discharge $\mathbf{q}$ is a vector quantity (denoted by **bold** type) that, in Cartesian coordinates, can be represented by

$$\mathbf{q} = (q_x, q_y, q_z) \tag{2.19}$$

or by the equivalent form

$$\mathbf{q} = q_x\hat{\mathbf{x}} + q_y\hat{\mathbf{y}} + q_z\hat{\mathbf{z}} \tag{2.20}$$

where $\hat{\mathbf{x}}$, $\hat{\mathbf{y}}$ and $\hat{\mathbf{z}}$ are unit vectors in the x, y and z directions, and q_x, q_y and q_z are projections of $\mathbf{q}$ on to those directions, respectively. Recall that each component of the vector $\mathbf{q}$ is, in itself, a three dimensional scalar function of space $q_u(x,y,z)$, where u is used here, and throughout the following text, to denote an arbitrary vector direction (in Cartesian coordinates, u = x, y or z).

Darcy's Law in Differential Form

If the hydraulic conductivity K should change, or if the hydraulic head h should be caused to vary in an arbitrary way, in a single coordinate direction, or along the axis of a particular experimental apparatus, it may be more appropriate to consider K or h as continuous functions of the relevant coordinate. To be specific, we assume that this is the direction of increasing x. For continuous, one dimensional, media, in the limit of small Δx, the flux in the x direction is

$$q_x = \lim_{\Delta x \to 0}\left(-K\frac{\Delta h}{\Delta x}\right) = -K\frac{dh}{dx} \tag{2.21}$$

or

$$q_x = -K\frac{dh}{dx} \tag{2.22}$$

This is Darcy's law in one dimensional differential form, and relates the specific discharge in the x direction q_x to the differential change of hydraulic head dh/dx in the x direction.

Clearly y or z can be substituted for x, if the flow geometry is in these directions. If flow can occur in all three dimensions simultaneously, the individual components are superimposed according to

$$\mathbf{q} = -\, \hat{\mathbf{x}}\, K\, \frac{\partial h}{\partial x} - \hat{\mathbf{y}}\, K\, \frac{\partial h}{\partial y} - \hat{\mathbf{z}}\, K\, \frac{\partial h}{\partial z} \qquad (2.23)$$

Recalling from vector calculus that the gradient operator is given by

$$\mathbf{grad} = \nabla = \hat{\mathbf{x}}\, \frac{\partial}{\partial x} + \hat{\mathbf{y}}\, \frac{\partial}{\partial y} + \hat{\mathbf{z}}\, \frac{\partial}{\partial z} \qquad (2.24)$$

we might relate the specific discharge vector $\mathbf{q}$ to the *hydraulic gradient* through

$$\mathbf{q} = -\, K\, \mathbf{grad}\, h \qquad (2.25)$$

or

$$\mathbf{q} = -\, K\, \nabla h \qquad (2.26)$$

INHOMOGENEOUS VERSUS ANISOTROPIC MEDIA

Types of Media

It is important to recognize the difference between the following terms.

Homogeneous versus inhomogeneous. A homogeneous or uniform medium is one whose properties are the same throughout its volume. If its properties vary from point to point, a medium is said to be inhomogeneous, nonuniform or heterogeneous. In the latter case K = K(x,y,z).

Isotropy versus anisotropy. An isotropic material has the same physical properties on a microscopic scale at a point, regardless of the direction in which measurements are made. An anisotropic material has microscopic properties that are directionally dependent. The latter case is discussed further below.

One should not equate homogeneity with isotropy. Some materials can be uniformly (or homogeneously) anisotropic, and other materials may be nonuniform (inhomogeneous) and isotropic. In general, isotropy (or anisotropy) is a property characterizing a medium at a point, whereas homogeneity (or inhomogeneity) is a term used to describe the larger scale property of a medium throughout its volume, such as an entire aquifer, or a geologic section. The simplest case to treat mathematically is that of material which is uniform and isotropic.

Anisotropy

Consider a situation where a hydraulic gradient applied in one direction across a material leads to a different flux than a gradient applied across an orthogonal direction. Assume that the following directional dependence of q, K and dh/dl applies:

$$q_x = - K_x \; \partial h/\partial x \qquad \text{(2.27a)}$$
$$q_y = - K_y \; \partial h/\partial y \qquad \text{(2.27b)}$$
$$q_z = - K_z \; \partial h/\partial z \qquad \text{(2.27c)}$$

The Hydraulic Conductivity Tensor

The above is a special case of the more general relationship:

$$q_x = - K_{xx} \; \partial h/\partial x - K_{xy} \; \partial h/\partial y - K_{xz} \; \partial h/\partial z \qquad \text{(2.28a)}$$
$$q_y = - K_{yx} \; \partial h/\partial x - K_{yy} \; \partial h/\partial y - K_{yz} \; \partial h/\partial z \qquad \text{(2.28b)}$$
$$q_z = - K_{zx} \; \partial h/\partial x - K_{zy} \; \partial h/\partial y - K_{zz} \; \partial h/\partial z \qquad \text{(2.28c)}$$

This may be seen by simply considering the implications of the expressions on an intuitive level.

Alternatively, we can develop the latter expressions through a simple rotation of coordinates of the original anisotropic form. To illustrate the latter approach, relation (2.27) in the original "principal coordinates" can be rewritten in the form of a matrix operation (or *tensor* product)

$$\mathbf{q} = - \mathbf{K} \, \nabla h \qquad \text{(2.29)}$$

where $\mathbf{q}$ and ∇h are column vectors, and $\mathbf{K}$ is the matrix (or tensor) given by

$$\mathbf{K} = \begin{bmatrix} K_x & 0 & 0 \\ 0 & K_y & 0 \\ 0 & 0 & K_z \end{bmatrix} \qquad \text{(2.30)}$$

Now consider the simple coordinate rotation $\mathbf{R}$, represented by the two operations

$$\mathbf{q'} = \mathbf{R} \; \mathbf{q} \qquad \text{(2.31)}$$

and

$$\nabla'h = \mathbf{R} \; \nabla h \qquad \text{(2.32)}$$

Thus

$$\mathbf{q'} = \mathbf{R} \; [\mathbf{K} \; [\mathbf{R}^{-1} \; \nabla'h] \;] \qquad \text{(2.33)}$$

which can be rewritten in the form

$$\mathbf{q'} = - \; \mathbf{K'} \; \nabla'h \qquad \text{(2.34)}$$

where

$$\mathbf{K'} = \mathbf{R} \; \mathbf{K} \; \mathbf{R}^{-1} \qquad \text{(2.35)}$$

This general form for the conductivity tensor can be "diagonalized" to the previous form of (2.30) through a suitable choice of coordinates, and if the material is isotropic, (2.30) simplifies further to (2.26), since its diagonal terms would be equal.

DIMENSIONALITY OF PRACTICAL HYDROGEOLOGICAL SITUATIONS: SIMPLIFYING A COMPLEX WORLD TO LOWER DIMENSIONS

A Three-Dimensional View

Hydrogeologists often couch their calculations in terms of one, two and three dimensional models. This section describes the rationale of such a procedure in terms of a specific example: the idealized hydrogeological situation is shown in Figure 2.4a.

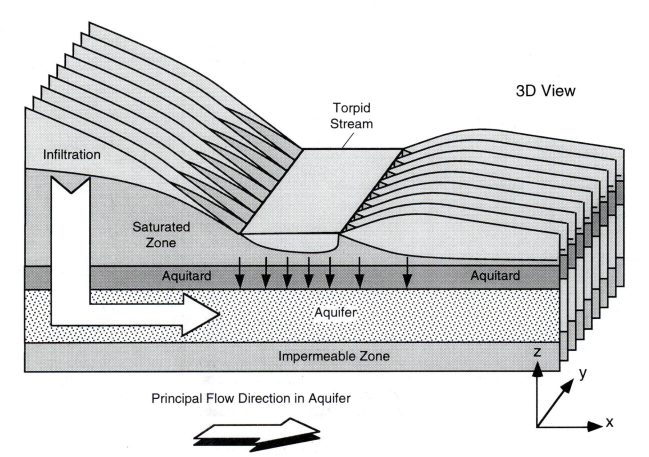

Figure 2.4a A vertical cutaway of an idealized 3-D earth model. The section is oriented along the principal groundwater flow direction. Note that, in this case, a series of parallel vertical sections perpendicular to the physiographic axis of the valley are quite similar in character.

This schematic represents infiltration of groundwater in the highlands on the left, causing regional recharge of a "leaky" confined aquifer. The principal direction of fluid flow in the aquifer is to the right, along which the x axis of a local cartesian coordinate system is oriented. The y axis of the coordinate system is parallel to the physiographic axis of the valley, along which a relatively static body of surface water lies. This might be a torpid stream, a lake, a reservoir or a saturated wetland.

A number of interesting questions are associated with such a situation, one of which is the interaction of groundwater flow with various recharge and discharge processes. For example, what is the effect of leakage from the torpid stream on regional groundwater flow in the confined aquifer? (Actually a *leaky* confined aquifer.)

A Two Dimensional Model

To address such a question in three-dimensions (3-D) is a tedious task at best, and often formidable. While in fact the world is 3-D, it is not uncommon for a set of parallel, vertical cross-sections to reveal similar geometries as illustrated in Figure 2.4a. While in any practical situation there is usually some variation between sections, however modest, in many cases such differences are far less significant than variations within each section. For the case shown here, variations are much stronger in the x-z plane than they are in the y-z plane, for example.

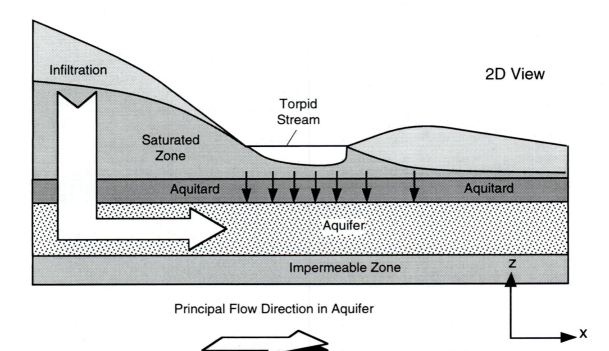

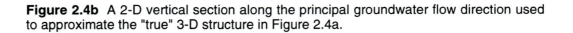

Figure 2.4b A 2-D vertical section along the principal groundwater flow direction used to approximate the "true" 3-D structure in Figure 2.4a.

One would expect much stronger flow, and variations in flow due to changing physical conditions, perpendicular to the physiographic axis of the valley, than along it. This condition may permit one to reasonably approximate the real 3-D earth by a simple two dimensional (2-D) vertical section as shown in Figure 2.4b.

Such a 2-D model might be quite adequate, for example, to address the question posed above: "What is the effect of leakage from the torpid stream on regional groundwater flow in the leaky confined aquifer?"

A One Dimensional Model

In some cases, significant questions can be still addressed by even a simpler class of models. For example, if one wants to address the specific question of how groundwater flow in the principal direction (assumed to be along the x axis) is modified by leakage from the torpid stream shown in Figure 2.4a, a simple one dimensional (1-D) model, such as shown in Figure 2.4c, might be quite sufficient.

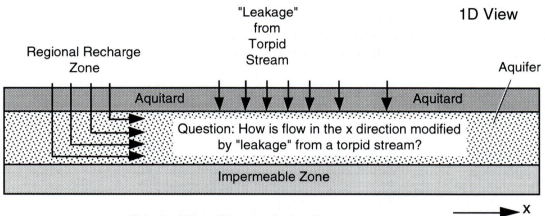

(Assume in x direction)

Figure 2.4c A 1-D representation of a portion of the 3-D situation shown in Figure 2.4a. Here the question is narrowly posed regarding the modification of the principal flow component by leakage from surface sources.

In such a case, one might assume that any vertical recharge or discharge from the aquifer can be represented as local or distributed sources or sinks of fluid along the axis of the 1-D "aquifer". This might be appropriate, for example, providing the vertical flux is distributed over the thickness of the aquifer "instantaneously" — that is to say in a time that is short compared to the horizontal transit time of the principal horizontal flow through the aquifer. In the next chapter we discuss ways to represent processes that in reality are multidimensional, yet can be reasonably approximated by models of lower dimensionality.

Chapter 3.   Taylor's Series, Directional Derivatives and Hydraulic Gradients

TAYLOR'S SERIES FOR HYDRAULIC HEAD

Variable Flow in One Dimension

Consider the case of variable flow along one axis. We assume that while there may be various sources and sinks of fluid along the path of flow, details as to how this fluid enters and leaves the system (and which might well require local three dimensional flow) are not important, only the fact that these sources and sinks modify an overall one dimensional flux. A hypothetical profile of the hydraulic head measured by standpipes or monitoring wells along the axis of this flow is shown in the lower panel.

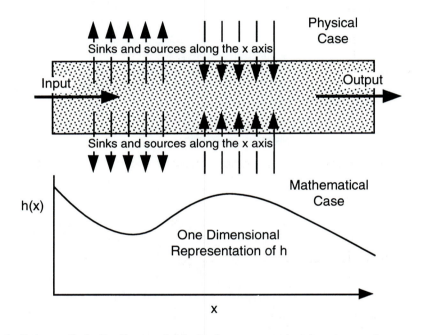

Figure 3.1 Schematic indicating variable h along one axis, showing various sources and sinks for the fluid along the path of flow

We are interested in characterizing the hydraulic head along the principal axis of the system. Thus h is represented as a 1-D function or $h(x)$.

Extrapolating the Value of a Function from One Point to Another

A frequent objective of mathematical analysis is, given complete information of a function at a given point x_o — that is to say $h(x_o)$ and all its necessary properties — extrapolate to the value of h at an arbitrary point x, or $h(x)$ as shown in Figure 3.2.

One way to do this is to use the tangent to the curve at x_o (point a) and to "project" the value of h at x_o to the value of h at x (point b) as illustrated in Figure 3.3.

Assuming the tangent to the curve at x_0 is given by the derivative of h at x_0 (denoted by $h'(x_0)$), an approximate value of $h(x)$ (at point b) is given by

$$h_{approx}(x) = h(x_0) + (x-x_0)\, h'(x_0) \tag{3.1}$$

which, as shown in Figure 3.4, usually works quite well if the distances $x-x_0$ are small, and curvature of the function is not great.

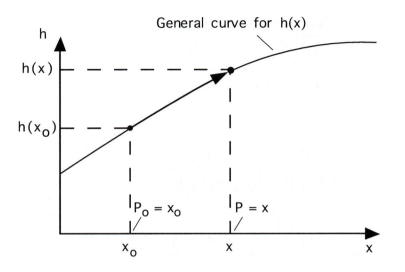

Figure 3.2 Extrapolating h to a new position x from an original value of h at x_0.

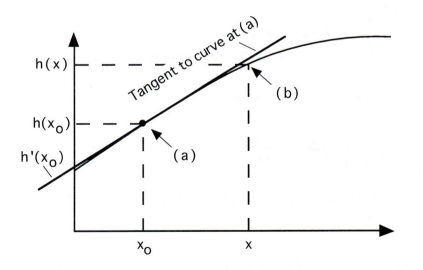

Figure 3.3 Showing the tangent to the curve at x_0 (point a) that can be used to "project" an approximation of h from x_0 to x (point b).

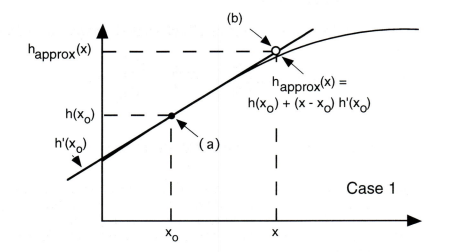

Figure 3.4 Employing the tangent to the curve at x_o (the first derivative of h at point a) to linearly extrapolate to an approximate value of h at x (point b).

Problems with Extrapolating over Large Distance

In the example in Figure 3.5 (Case 2), the extrapolation is over too large a distance to lead to accurate results, considering the curvature of the function.

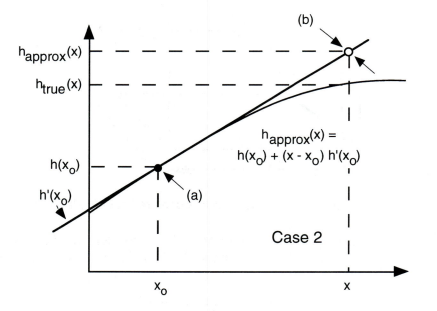

Figure 3.5 Illustrating the growing inaccuracy of "projecting" h(x) over large distances using linear extrapolation.

Since from elementary calculus, we will recall that the "curvature" of a function is related to its second derivative, our inadequate extrapolation in Figure 3.5 implies that we need to take into account higher order derivatives in our extrapolation. Such a "higher order" extrapolation might be of the form:

$$h(x) = h(x_o) + (x - x_o) h'(x_o) + [(x - x_o)^2] h''(x_o) / 2 + \dots \qquad (3.2)$$

Taylor's Series Expansion of a Function h(x)

These considerations lead naturally to the concept of a Taylor's Series expansion of a function:

$$h(x) = h(x_0) + (x-x_0)\frac{\partial h}{\partial x}\Big|_{x_0} + \frac{(x-x_0)^2}{2}\frac{\partial^2 h}{\partial x^2}\Big|_{x_0} + \ldots \qquad (3.3)$$

TAYLOR'S SERIES IN THREE DIMENSIONS

Our objective is to describe the behavior of groundwater flow in a small (infinitesimal) region of space, and relate this to statements of physical relations at a "point". To do so, we begin by considering a rectilinear elemental reference volume having sides of dimensions Δx, Δy, and Δz, respectively, and a Cartesian coordinate system having a local origin at point $P_0 = (x_0, y_0, z_0)$ at the center of the volume as shown in Figure 3.6.

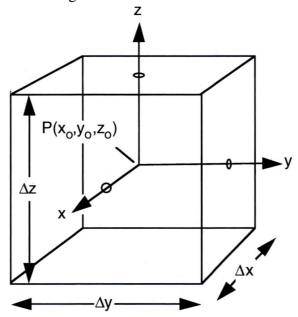

Figure 3.6 An elemental rectilinear volume with a centered coordinate system.

Consider the Cartesian geometry illustrated in Figure 3.7. A line element δl is directed from the local origin at $P_0 = (x_0, y_0, z_0)$ to an arbitrary point at $P = (x, y, z)$. In terms of its Cartesian coordinates, δl is given by

$$\delta l = \hat{x}(x - x_0) + \hat{y}(y - y_0) + \hat{z}(z - z_0) \qquad (3.4)$$

or

$$\delta l = \hat{x}\,\delta x + \hat{y}\,\delta y + \hat{z}\,\delta z \qquad (3.5)$$

where we denote incremental distances, or offsets, by

$$\delta x = x - x_0 \qquad (3.6a)$$
$$\delta y = y - y_0 \qquad (3.6b)$$
$$\delta z = z - z_0 \qquad (3.6c)$$

Note: We will use the notation, throughout the text, of a lower case Greek delta (δ) to denote a small, but finite, variable offset in a coordinate position or functional value. If a small, but finite, fixed distance is considered, we will use the notation of an upper case Greek delta (Δ). (Partial derivatives will be denoted with Roman lower case deltas, ∂.)

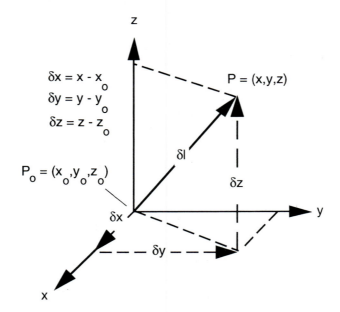

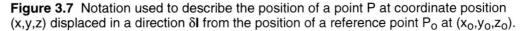

Figure 3.7 Notation used to describe the position of a point P at coordinate position (x,y,z) displaced in a direction δl from the position of a reference point P_0 at (x_0,y_0,z_0).

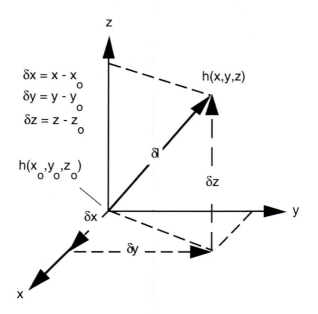

Figure 3.8 Notation used to describe the relation between the value of the function h (x,y,z) at the point P = (x,y,z) displaced in a direction δl from the reference point $P_0 = (x_0,y_0,z_0)$, where the function has the value h (x_0,y_0,z_0).

The objective of our discussion is to explore the nature of the function $h(x,y,z)$ at the point $P = (x,y,z)$ relative to its value at $P_o = (x_o, y_o, z_o)$, as shown in Figure 3.8. To do so, we expand the function $h(x,y,z)$ at a point $P = (x,y,z)$ as a Taylor's Series relative to its value at $P_o = (x_o, y_o, z_o)$. We thus obtain

$$h(x,y,z) = h(x_o, y_o, z_o) + \delta x \left.\frac{\partial h}{\partial x}\right|_{P_o} + \delta y \left.\frac{\partial h}{\partial y}\right|_{P_o} + \delta z \left.\frac{\partial h}{\partial z}\right|_{P_o} +$$

$$+ \frac{\delta x^2}{2} \left.\frac{\partial^2 h}{\partial x^2}\right|_{P_o} + \frac{\delta y^2}{2} \left.\frac{\partial^2 h}{\partial y^2}\right|_{P_o} + \frac{\delta z^2}{2} \left.\frac{\partial^2 h}{\partial z^2}\right|_{P_o} +$$

$$+ \delta x \delta y \left.\frac{\partial^2 h}{\partial x \partial y}\right|_{P_o} + \delta x \delta z \left.\frac{\partial^2 h}{\partial x \partial z}\right|_{P_o} + \delta y \delta z \left.\frac{\partial^2 h}{\partial y \partial z}\right|_{P_o} +$$

$$+ \delta x \delta y \delta z \left.\frac{\partial^3 h}{\partial x \partial y \partial z}\right|_{P_o} + ... + R \left[\delta u^n \left.\frac{\partial^n h}{\partial u^n}\right|_{P_o} \right] \qquad (3.7)$$

where, it is important to remember, all derivatives on the right hand side are evaluated at the local origin P_o, *not* at the position at which $h(x,y,z)$ is being evaluated. The remainder, upon truncation of the Taylor's series, is represented by the term

$$R \left[\delta u^n \left.\frac{\partial^n h}{\partial u^n}\right|_{P_o} \right] \qquad (3.8)$$

where the variable "u" is used to denote one of the Cartesian coordinates: x, y or z. There are, of course, an infinite number of terms involving a variety of cross-derivatives. In truncating a Taylor's series at a certain order (often after the first), one assumes that the function is sufficiently "well-behaved" such that either the higher order derivatives are negligible, or that the displacement coefficients, δu^n, are small; or better yet, both conditions apply.

Directional Derivatives of Hydraulic Head

In describing subsurface flow, we are often interested in characterizing the *difference* between hydraulic heads in a specific direction.

$$h(x,y,z) - h(x_o,y_o,z_o) = \delta x \left.\frac{\partial h}{\partial x}\right|_{P_o} + \delta y \left.\frac{\partial h}{\partial y}\right|_{P_o} + \delta z \left.\frac{\partial h}{\partial z}\right|_{P_o} +$$

$$+ \frac{\delta x^2}{2} \left.\frac{\partial^2 h}{\partial x^2}\right|_{P_o} + \frac{\delta y^2}{2} \left.\frac{\partial^2 h}{\partial y^2}\right|_{P_o} + \frac{\delta z^2}{2} \left.\frac{\partial^2 h}{\partial z^2}\right|_{P_o} +$$

$$+ \delta x \delta y \left.\frac{\partial^2 h}{\partial x \partial y}\right|_{P_o} + \delta x \delta z \left.\frac{\partial^2 h}{\partial x \partial z}\right|_{P_o} + \delta y \delta z \left.\frac{\partial^2 h}{\partial y \partial z}\right|_{P_o} +$$

$$+ \delta x \delta y \delta z \left.\frac{\partial^3 h}{\partial x \partial y \partial z}\right|_{P_o} + ... + R\left[\delta u^n \left.\frac{\partial^n h}{\partial u^n}\right|_{P_o}\right] \qquad (3.9)$$

Let

$$\delta h = h - h_o = h(x,y,z) - h(x_o,y_o,z_o) \qquad (3.10)$$

We can define a "directional difference" (or more precisely a directional "finite" difference) as

$$\frac{\delta h}{\delta l} = \frac{h - h_o}{\delta l} \qquad (3.11)$$

or

$$\frac{\delta h}{\delta l} = \frac{h - h_o}{\delta l} = \frac{\delta x}{\delta l} \left.\frac{\partial h}{\partial x}\right|_{P_o} + \frac{\delta y}{\delta l} \left.\frac{\partial h}{\partial y}\right|_{P_o} + \frac{\delta z}{\delta l} \left.\frac{\partial h}{\partial z}\right|_{P_o} +$$

$$+ R\left[\frac{\delta u^n}{\delta l} \left.\frac{\partial^n h}{\partial u^n}\right|_{P_o}\right] \qquad (3.12)$$

Direction Cosines

The latter (3.12) can be written as

$$\frac{\delta h}{\delta l} = \frac{h - h_o}{\delta l} = \cos \alpha \left.\frac{\partial h}{\partial x}\right|_{P_o} + \cos \beta \left.\frac{\partial h}{\partial y}\right|_{P_o} + \cos \gamma \left.\frac{\partial h}{\partial z}\right|_{P_o} +$$

$$+ R\left[\frac{\delta u^n}{\delta l} \left.\frac{\partial^n h}{\partial u^n}\right|_{P_o}\right] \qquad (3.13)$$

where $\cos \alpha$, $\cos \beta$ and $\cos \gamma$ are the direction cosines, as shown in Figure 3.9, between the direction of the directed line element $\delta \mathbf{l}$ and the x, y and z axes, respectively.

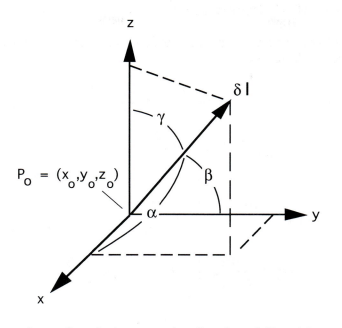

Figure 3.9 The angles α, β and γ between the direction of $\delta \mathbf{l}$ and the x, y and z axes, respectively.

In the limit of $\delta \mathbf{l}$ becoming small, the left hand term becomes a partial derivative

$$\frac{\partial h}{\partial l} = \lim_{\delta l \to 0} \frac{\delta h}{\delta l} \qquad (3.14)$$

and, upon neglecting higher order terms in δu^n on the right hand side of the relation, the previous expression becomes

$$\frac{\partial h}{\partial l} = \nabla h \cdot \hat{\mathbf{l}} \qquad (3.15)$$

where

$$\nabla h = \hat{\mathbf{x}} \frac{\partial h}{\partial x} + \hat{\mathbf{y}} \frac{\partial h}{\partial y} + \hat{\mathbf{z}} \frac{\partial h}{\partial z} \qquad (3.16)$$

and

$$\hat{\mathbf{l}} = \hat{\mathbf{x}} \cos \alpha + \hat{\mathbf{y}} \cos \beta + \hat{\mathbf{z}} \cos \gamma \qquad (3.17)$$

Principal Directions, Hydraulic Gradients and Flow-Lines

Hydraulic gradient. That principal direction for which $|\partial h/\partial l|$ is maximum, and h is increasing in amplitude, is often referred to as the "direction of the hydraulic gradient", and the magnitude of $\partial h/\partial l$ in that direction is termed the hydraulic gradient. An example of elevation contours of the hydraulic head is shown in Figure 3.10, along with the associated hydraulic gradient vectors (in plan view) at selected points. Note that the gradient vectors point in the direction of increasing elevation, and their magnitude is proportional to |∂h/∂l|. The opposite direction is well known to skiers as the "fall line", and is the quickest way back to the base lodge.

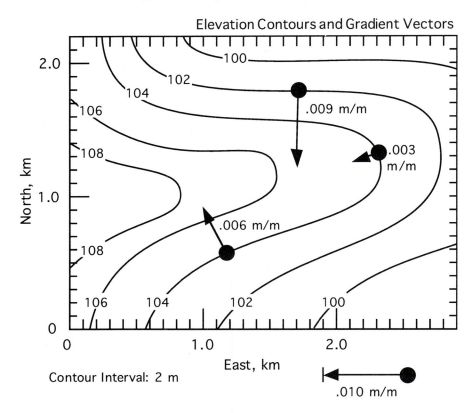

Figure 3.10 Elevation contours of hydraulic head and associated gradient vectors at selected points. The direction of each vector corresponds to the direction of the most rapid rate of *increase* of h with distance, and the magnitude is proportional to the magnitude of |∂h/∂l| in that direction. The units for the gradient used here are m/m; often one uses m/km.

Fluid flow. From Darcy's law, the direction of maximum flow will be in the direction of the maximum rate of *decrease* of h, or in the direction of *negative* hydraulic gradient (the "fall line" from our skiing analogy). The magnitude of flow is proportional to the local value of hydraulic conductivity, so that the specific discharge at a point is a vector given by the product of hydraulic conductivity and the negative hydraulic gradient, or

$$\mathbf{q} = - K \, \nabla h \qquad (3.18)$$

With reference to Figure 4.10, local flow will be in a direction opposite to that of the arrows (because of the *negative* gradient).

Flow-lines. Steady-state flow in a source-free, homogeneous, isotropic medium, will be along a local "flow-line", and will cross orthogonally (perpendicularly) to lines (or surfaces) of h = constant, in a direction of *decreasing* head. An example showing the relation between flow-lines and h-lines (a flow-net) is shown in Figure 3.11.

Such flow nets are discussed more extensively later in the text.

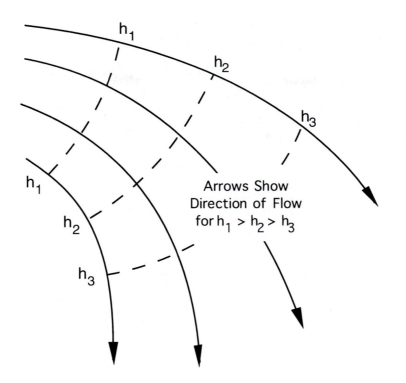

Figure 3.11 A simple two dimensional example illustrating fluid flow in the direction of negative hydraulic gradient (h decreases from h_1 to h_3).

Chapter 4. Conservation of Fluid Flow: Application to One Dimension

A CONTINUITY CONDITION ON DARCY FLOW

Assume that fluid flow has stabilized in time such that the net outward (positive) flux of fluid from a finite volume equals the rate at which fluid is produced by sources within the volume. Mathematically this can be stated as

$$\oiint_{\text{Surface}} \mathbf{q} \cdot d\mathbf{A} = \iiint_{\text{Vol}} [\text{All sources}] \, dV \tag{4.1}$$

or, if we use "w" for "water" to denote the source term on the right hand side of (4.1), we have

$$\oiint_{\text{Surface}} \mathbf{q} \cdot d\mathbf{A} = \iiint_{\text{Vol}} w(x,y,z) \, dV \tag{4.2}$$

Such relations where the flux through a closed surface is related to sources within the volume contained inside the surface are known as a "Conservation Laws". Note that, if there are no source terms, then $w(x,y,z) = 0$ everywhere within the volume, the right hand side of (4.2) is zero, and the relation reduces to

$$\oiint_{\text{Surface}} \mathbf{q} \cdot d\mathbf{A} = 0 \tag{4.3}$$

It should be emphasized that the latter expression does not indicate that the flux, $\mathbf{q}$, is zero; only that the net flux through the surface of the volume averages out to zero — whatever flows into the volume, flows out.

•   •   •

Aside: Relation to other fields of science

Conservation Laws have ubiquitous applications throughout physics, engineering and applied mathematics. In electromagnetic theory, the embodiment of such a relation is known as Gauss' Theorem, given by

$$\oiint_{\text{Surface}} \mathbf{D} \cdot d\mathbf{A} = \iiint_{\text{Vol}} \rho(x,y,z) \, dV \tag{a}$$

where, in SI units, $\mathbf{D}$ is the electric displacement field (or electric flux density) in coulombs m^{-2}, $\rho(x,y,z)$ is the electric charge density in coulombs m^{-3}.

•   •   •

Applying the Conservation Relation to One Dimension

We apply this relation to the simple one dimensional case in Figure 4.1, where flow is directed purely along the x axis, but where sources and sinks of fluid may exist along the axis. At some arbitrary point x_o along the axis of this flow, we center an elemental mathematical volume having a total length Δx. We assume x increases to the right, so that positive q_x is directed to the right. We accordingly term the left and right faces of this elemental volume, the "rear" and "forward" faces, respectively. The forward face is located at $x = x_o + \Delta x/2$, and the rear face is located at $x = x_o - \Delta x/2$. We now consider the behavior of the flow field throughout the elementary volume in more detail.

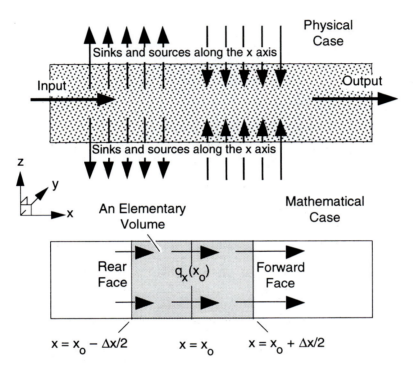

Figure 4.1 An elementary reference volume in 1-D for analyzing the conservation of flux with sources.

Assume that the flux or the specific discharge q_x is restricted to flow along the x axis, but because of local sources (inputs) or sinks (outputs), the intensity of q_x varies as a function of x. Moreover, assume that in the vicinity of any reference point x_o the flux q_x, along with its derivatives, is continuous. This would allow one to expand q_x in terms of a Taylor's series of the form:

$$q_x(x) = q_x(x_o) + (x-x_o)\frac{\partial q_x}{\partial x}\bigg|_{x_o} + \frac{(x-x_o)^2}{2}\frac{\partial^2 q_x}{\partial x^2}\bigg|_{x_o} + \dots \qquad (4.4)$$

Just as in the case of the Taylor's series expansion of $h(x)$ considered above, the Taylor's series of q_x allows one to project or extrapolate q_x from its value at x_o, to some distance away, for example to $x = x_o + \delta x$, where in (4.4),

$$\delta x = x - x_o \qquad (4.5)$$

Consider the particular case of projecting q_x to the boundaries of our one dimensional mathematical volume. The boundaries are located to the right and to the left of x_o, respectively, at the x positions

$$x_+ = x_o + \Delta x/2 \qquad (4.6a)$$

$$x_- = x_o - \Delta x/2 \qquad (4.6b)$$

as shown in Figures 4.1 and 4.2.

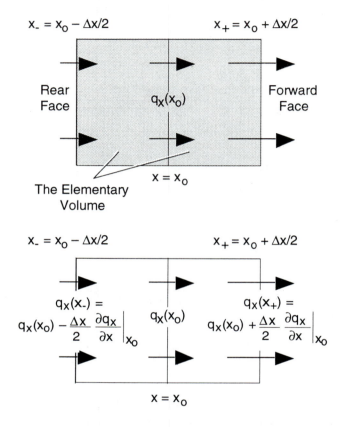

Figure 4.2 The Taylor's series form of q_x at the forward face and the rear face of the reference volume.

Assuming that the distance δx is sufficiently small so that the curvature of q_x can be neglected, we can neglect second and higher order terms in the above Taylor's series. Thus, Figure 4.2 illustrates that q_x can be represented along the right boundary and left boundary, respectively, of the reference volume by

$$q_x(x_+) \approx q_x(x_o) + \left(\frac{\Delta x}{2}\right)\frac{\partial q_x}{\partial x}\bigg|_{x_o} \qquad (4.7a)$$

and

$$q_x(x_-) \approx q_x(x_o) - \left(\frac{\Delta x}{2}\right)\frac{\partial q_x}{\partial x}\bigg|_{x_o} \qquad (4.7b)$$

Net Flux: Integrating the Left-Hand Side of the Conservation Relation

We next want to integrate the surface flux q_x across forward and rear interfaces, respectively. Because q_x and its derivative $\partial q_x / \partial x$, as used in the Taylor's series expansion, are their fixed values determined at $x = x_o$, and Δx is a constant of the problem, the corresponding values of $q_x(x_+)$ and $q_x(x_-)$ are constant over the surface of the respective interfaces at $x = x_+$ and $x = x_-$. Thus, the surface integrals over either side of the volume reduce to two contributions. Over the forward side, having a cross-sectional area (in the y-z plane) denoted by "AREA", the surface integral becomes

$$\iint_{\substack{\text{FORWARD} \\ \text{SIDE}}} q_x(x_+)\,\widehat{\mathbf{x}} \cdot d\mathbf{A} = q_x(x_+) \iint_{\substack{\text{FORWARD} \\ \text{SIDE}}} \widehat{\mathbf{x}} \cdot d\mathbf{A} \qquad (4.8)$$

where we have taken $q_x(x_+)$ out of the integral operation, because the flux is constant over the surface area perpendicular to flow at the position $x = x_o + \Delta x / 2$.

$$\iint_{\substack{\text{FORWARD} \\ \text{SIDE}}} \widehat{\mathbf{x}} \cdot d\mathbf{A} = + \text{AREA} \qquad (4.9)$$

$$q_x(x_+) \iint_{\substack{\text{FORWARD} \\ \text{SIDE}}} \widehat{\mathbf{x}} \cdot d\mathbf{A} = + q_x(x_+) \cdot \text{AREA} \qquad (4.10)$$

$$\iint_{\substack{\text{FORWARD} \\ \text{SIDE}}} q_x(x_+)\,\widehat{\mathbf{x}} \cdot d\mathbf{A} = + \text{AREA} \cdot q_x(x_+)$$

$$= + \text{AREA} \cdot \left[q_x(x_o) + \frac{\Delta x}{2} \left. \frac{\partial q_x}{\partial x} \right|_{x_o} \right] \qquad (4.11)$$

We may now apply this same set of operations to the rear surface of the reference volume. We must first note that through the usual conventions of vector calculus, the vector element of differential area $d\mathbf{A}$, because it an outward directed normal to the surface, points in the negative x direction. Thus, the relevant partial surface integral

$$\iint_{\substack{\text{REAR} \\ \text{SIDE}}} \widehat{\mathbf{x}} \cdot d\mathbf{A} = - \text{AREA} \qquad (4.12)$$

is negative for the rear surface. Thus, the net flux through the rear surface is given by

$$\iint_{\substack{\text{REAR} \\ \text{SIDE}}} q_x(x_-)\,\widehat{\mathbf{x}} \cdot d\mathbf{A} = - \text{AREA} \cdot q_x(x_-)$$

$$= - \text{AREA} \cdot \left[q_x(x_o) - \frac{\Delta x}{2} \left. \frac{\partial q_x}{\partial x} \right|_{x_o} \right] \qquad (4.13)$$

These two contributions — the net flux through the rear surface and the forward surface, respectively — are illustrated schematically in the Figure 4.3.

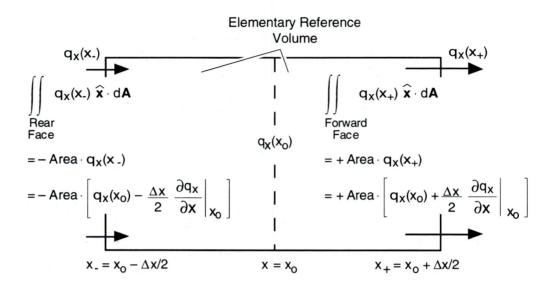

Figure 4.3 The surface integral of q_x over each face of the elementary 1-D reference volume, where the flux at each face is represented by a Taylor's series expansion at the mid-point $x = x_0$.

Thus, for this small, elemental volume, the left hand side (LHS) of the conservation equation (entailing the net flux through both the rear and forward surfaces, respectively) becomes

$$
\oiint_{Surface} \mathbf{q} \cdot d\mathbf{A} = \iint_{\substack{FORWARD \\ SIDE}} q_x(x_+)\hat{\mathbf{x}} \cdot d\mathbf{A} + \iint_{\substack{REAR \\ SIDE}} q_x(x_-)\hat{\mathbf{x}} \cdot d\mathbf{A}
$$

$$
= +AREA \cdot q_x(x_+) - AREA \cdot q_x(x_-)
$$

$$
= +AREA \cdot \left[q_x(x_0) + \frac{\Delta x}{2} \left.\frac{\partial q_x}{\partial x}\right|_{x_0} - q_x(x_0) + \frac{\Delta x}{2} \left.\frac{\partial q_x}{\partial x}\right|_{x_0} \right]
$$

$$
= +AREA \cdot \Delta x \left.\frac{\partial q_x}{\partial x}\right|_{x_0} \tag{4.14}
$$

Source Terms: Integrating the Right Hand Side of the Conservation Relation

We next consider the right hand side (RHS) of the conservation equation; the volume integral over all source terms. We first expand the source term $w(x)$ as a first order Taylor's series, such that

$$
w(x) = w(x_0) + (x - x_0)\left.\frac{\partial w}{\partial x}\right|_{x_0} \tag{4.15}
$$

Integrating $w(x)$ over the volume, we have

$$\iiint_{Vol} [\text{All sources}] \, dV = \iiint_{Vol} w(x) \, dxdydz \qquad (4.16)$$

Assuming $w(x)$ is constant over a uniform cross-sectional area in the y and z directions, the order of integration can be rearranged to the form

$$\iiint_{Vol} w(x) \, dxdydz = \int_{x_-}^{x_+} w(x) \left[\iint_{\substack{CROSS \\ SECTION}} dydz \right] dx \qquad (4.17)$$

where, since

$$\iint_{\substack{CROSS \\ SECTION}} dydz = + \text{AREA} \qquad (4.18)$$

we obtain

$$\iiint_{Vol} w(x) \, dxdydz = \text{AREA} \cdot \int_{x_-}^{x_+} w(x) \, dx \qquad (4.19)$$

Using Taylor' series to expand the argument of the integral over x leads to

$$\text{AREA} \cdot \int_{x_-}^{x_+} w(x) \, dx = \text{AREA} \cdot \int_{x_0 - \frac{\Delta x}{2}}^{x_0 + \frac{\Delta x}{2}} \left[w(x_0) + (x - x_0) \left. \frac{\partial w}{\partial x} \right|_{x_0} \right] dx \qquad (4.20)$$

which integrates out to

$$= + \text{AREA} \cdot \Delta x \cdot w(x_0) \qquad (4.21)$$

In short, the right hand volume integral over all source terms in our conservation relation leads to

$$\iiint_{Vol} [\text{All sources}] \, dV = + \text{AREA} \cdot \Delta x \cdot w(x_0) \qquad (4.22)$$

Balancing the Continuity Equation

Equating the left hand side (LHS) to the right hand side (RHS) of the continuity equation (4.1), employing (4.14) and (4.22), respectively, leads to

$$\oiint_{\text{Surface}} \mathbf{q} \cdot d\mathbf{A} = \iiint_{\text{Vol}} [\text{All sources}]\, dV$$

$$+\text{AREA} \cdot \Delta x \cdot \left.\frac{\partial q_x}{\partial x}\right|_{x_o} = +\text{AREA} \cdot \Delta x \cdot w(x_o)$$

$$\Delta V \left.\frac{\partial q_x}{\partial x}\right|_{x_o} = \Delta V\, w(x_o) \tag{4.23}$$

where the differential volume ΔV of our reference element is given by

$$\Delta V = \text{AREA}\ \Delta x \tag{4.24}$$

Cancelling the differential volume term from (4.23), we obtain

$$\left.\frac{\partial q_x}{\partial x}\right|_{x_o} = w(x_o) \tag{4.25}$$

Since our reference point x_o was quite arbitrary along the x axis, the expression must hold for any x, so that

$$\frac{\partial q_x}{\partial x} = w(x) \tag{4.26}$$

This expression is known as a *one dimensional divergence relation*.

One Dimensional Flow

Substituting the right hand side of the following one-dimensional form of Darcy's law

$$q_x = -K\frac{dh}{dx} \tag{4.27}$$

for q_x in (4.26), we obtain

$$\frac{\partial}{\partial x}\left[K(x)\frac{\partial h}{\partial x}\right] = -w(x) \tag{4.28}$$

But if K is constant (i.e. the medium is homogeneous and isotropic), we have

$$\frac{\partial^2 h}{\partial x^2} = -\frac{w}{K} \tag{4.29}$$

which, since the second derivative of h is negative, suggests that the functional form of h is concave down. This is certainly in keeping with our physical concepts of the source term w "driving" the hydraulic head h as we see from the following.

If we introduce infiltration into a static situation of uniform hydraulic head, the new steady-state configuration of the piezometric surface is shown in Figure 4.4.

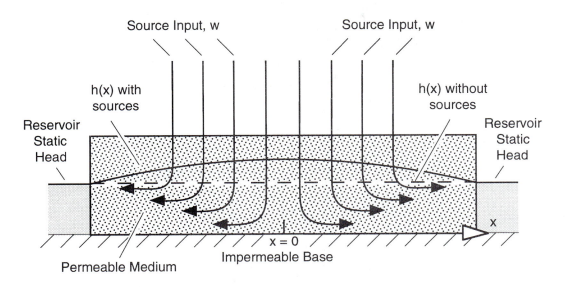

Figure 4.4 A 1-D model of hydraulic head in the presence of infiltration. The left and right margins of the permeable zone are maintained at a constant static head by reservoirs.

The shape of the piezometric surface is symmetrical about the midpoint (x=0), with the magnitude of the slope increasing in either direction away from the midpoint, implying that the magnitude of the flux (since $|q_x| \propto |\partial h/\partial x|$) is monotonically increasing in either direction as well. Obviously, at the midpoint, because of symmetry, the slope of the surface is zero, or $\partial h/\partial x = 0$ (and $q_x = 0$). This in turn implies from Darcy's law, that, at the midpoint, $q_x = -K\partial h/\partial x = 0$. To the left of the midpoint, the slope is increasingly positive (i.e. h(x) increases to the right with increasing x), which from Darcy's law implies an increasingly negative flux, q_x (i.e. the flux is directed to the left). To the right of the midpoint, the slope is increasingly negative, which from Darcy's law implies an increasingly positive flux, q_x (i.e. the flux is directed to the right)

Source-Free Conditions

In a source-free region, w = 0, so that

$$\frac{\partial^2 h}{\partial x^2} = 0 \tag{4.30}$$

which would imply that the curvature of a function h(x) is zero in a source-free region. This would imply that, in such a case, the gradient ($\partial h/\partial x$) would be constant, and, thus, the specific discharge q_x would be constant.

APPLICATION TO 1-D FLOW IN A CONFINED AQUIFER

Laplace's Equation Applied to 1-D Flow

Consider the geometry illustrated in Figure 4.5. (Note: This is a simplified counterpart to the problem of an unconfined aquifer with infiltration considered in detail by Fetter, 1994.) Assume we have a vertical 2-D section cutting through two water-filled channels or canals separated a distance L. The hydraulic head of each channel is constant in time. The height of water in Channel #1 is maintained at a constant head of h_1, and the height of water in Channel #2 is maintained at a constant head of h_2. An aquifer of uniform thickness b and uniform hydraulic conductivity K connects the two. The aquifer is bounded above and below by impermeable confining layers. All flow, therefore, is confined to be strictly in the horizontal x direction.

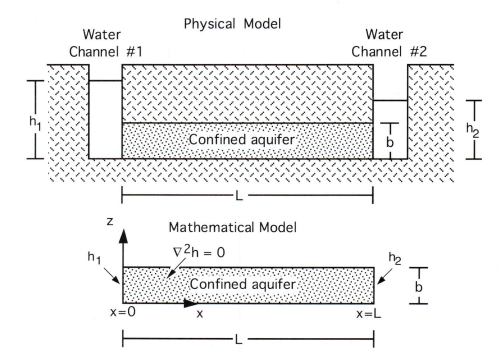

Figure 4.5 Example of a 1-D confined aquifer without sources. The physical model is shown in the top panel, whereas the mathematical model is shown in the bottom.

For the specific case considered here, we have to solve the 1-D Laplace equation

$$d^2h/dx^2 = 0 \qquad (4.31)$$

subject to the boundary conditions that $h = h_1$ at $x = 0$, and $h = h_2$ at $x = L$.

Some Preliminary Considerations

Having determined what we feel is the proper differential form to "solve", before we proceed too far along in our derivation, it is always appropriate to review the consistency of the differential form with the underlying assumptions and the fundamental relations that govern the physics of the problem.

The most fundamental relation is Darcy's law, which in the 1-D case considered here should have the form:

$$q_x = -K \, dh/dx \tag{4.32}$$

where, according to Figure 4.5, the direction of flow is clearly from the higher hydraulic head on the left to the lower hydraulic head on the right.

We define the "total discharge" Q_x as the total flux integrated over the thickness of the aquifer, where in the present case

$$Q_x = q_x b \tag{4.33}$$

where q_x is the specific discharge, and b is the thickness of the aquifer.

We can also define the "*transmissivity*" of the aquifer as the conductivity integrated over the total thickness of the aquifer, which in the present case, since K is uniform, is simply the conductivity-thickness product given by

$$T = Kb \tag{4.34}$$

If we multiply both sides of (4.32) by the thickness of the aquifer b, the resulting form can be expressed in terms of the latter two parameters as

$$Q_x = -T \, dh/dx \tag{4.35}$$

where Q_x is the total discharge throughout the entire thickness of an aquifer having a transmissivity of T.

Conservation of the total flux (the continuity condition on Q_x) would indicate that whatever fluid flows into the aquifer on the left flows out of the aquifer on the right. This is true throughout the length of the aquifer, so that from the last section we have

$$dQ_x / dx = 0 \tag{4.36}$$

which is a differential equation whose solution is that Q_x is constant with x, or

$$Q_x = C \tag{4.37}$$

This in turn implies from (4.35) that

$$dh/dx = -C/T \tag{4.38}$$

or that h is linear (constant "slope" or gradient) with x.

We could integrate (4.38) directly, subject to the prescribed boundary conditions at each end of the aquifer, but at this point we will return to the second derivative form provided by Laplace's equation (4.31).

Solving Laplace's Equation as a Boundary Value Problem in 1-D

We proceed to solve the 1-D Laplace equation

$$d^2 h / dx^2 = 0 \tag{4.31}$$

subject to the boundary conditions that $h = h_1$ at $x = 0$, and $h = h_2$ at $x = L$. Writing the equation in the form

$$d[dh/dx]/dx = 0 \qquad (4.39)$$

we can integrate once to obtain

$$dh/dx = \text{a constant} \qquad (4.40)$$

We denote the constant by A, and now integrate

$$dh/dx = A \qquad (4.41)$$

to give

$$h(x) = Ax + B \qquad (4.42)$$

where B is simply a second constant of integration. The values of A and B are determined from the boundary conditions on h at $x = 0$ and $x = L$. At $x = 0$

$$h(0) = h_1 \qquad (4.43)$$

or

$$h(0) = A \cdot 0 + B \qquad (4.44)$$

which implies

$$B = h_1 \qquad (4.45)$$

At $x = L$

$$h(L) = h_2 = A L + h_1 \qquad (4.46)$$

which can be solved for

$$A = [h_2 - h_1]/L \qquad (4.47)$$

Thus, our general solution (4.42) can be rewritten in terms of the above expressions for A and B to determine the following equation for the hydraulic head:

$$h(x) = [(h_2 - h_1)/L] x + h_1 \qquad (4.48)$$

which implies that $h(x)$ is linear with x

Employing Darcy's law, (4.48) can be differentiated with respect to x to yield the following expression for the total discharge:

$$Q_x = - T [(h_2 - h_1)/L] \qquad (4.49)$$

As we expected from our preliminary considerations above, Q_x is constant along the length of the aquifer.

Chapter 5. Fundamental Relations for Groundwater Flow in Three Dimensions

SYNOPSIS OF BASIC FORMS

Flow in Three Dimensional Distributed Media

Recall that by distributed media, we mean those cases where material properties and flow conditions are not strictly controlled in a contained vessel under laboratory conditions. Intrinsic to the concept of distributed systems is that, usually, the properties of the medium and/or the flow parameters vary from point to point in one, two or three spatial dimensions; as well as, perhaps, in time. In general, the hydraulic conductivity is a scalar function of three spatial dimensions, such that

$$K = K(x,y,z) \tag{5.1}$$

and might have a time dependence as well, so that, in certain cases, $K = K(x,y,z,t)$, but we will not treat the latter case in this text.

In the most general case, fluid flow is in three dimensions, so that the specific discharge $\mathbf{q}$ at a point $P = (x,y,z)$ is a vector quantity (denoted by **bold** type) that, in Cartesian coordinates, can be represented by a triplet of values

$$\mathbf{q}(x,y,z) = (q_x(x,y,z), q_y(x,y,z), q_z(x,y,z)) \tag{5.2}$$

or by the equivalent form

$$\mathbf{q}(x,y,z) = q_x(x,y,z)\widehat{\mathbf{x}} + q_y(x,y,z)\widehat{\mathbf{y}} + q_z(x,y,z)\widehat{\mathbf{z}} \tag{5.3}$$

where $\widehat{\mathbf{x}}$, $\widehat{\mathbf{y}}$ and $\widehat{\mathbf{z}}$ are unit vectors in the x, y and z directions, and $q_x(x,y,z)$, $q_y(x,y,z)$ and $q_z(x,y,z)$ are projections of $\mathbf{q}$ on to those directions, respectively. Each component of the vector $\mathbf{q}$ is a three dimensional scalar function of space $q_u(x,y,z)$, where, as earlier in our discussion, u is used to denote an arbitrary vector component. In Cartesian coordinates, u = x, y or z. With reference to the geometry in Figure 5.1, if each of the components q_u is sufficiently continuous at a point $P_o = (x_o,y_o,z_o)$, $q_u(x,y,z)$ can be expanded as a Taylor's series in the vicinity of its value at $q_u(x_o,y_o,z_o)$ according to

$$q_u(x,y,z) = q_u(x_o,y_o,z_o) + \delta x \left.\frac{\partial q_u}{\partial x}\right|_{P_o} + \delta y \left.\frac{\partial q_u}{\partial y}\right|_{P_o} + \delta z \left.\frac{\partial q_u}{\partial z}\right|_{P_o} + \frac{\delta x^2}{2}\left.\frac{\partial^2 q_u}{\partial x^2}\right|_{P_o}$$

$$+ \frac{\delta y^2}{2}\left.\frac{\partial^2 q_u}{\partial y^2}\right|_{P_o} + \frac{\delta z^2}{2}\left.\frac{\partial^2 q_u}{\partial z^2}\right|_{P_o} + \delta x \delta y \left.\frac{\partial^2 q_u}{\partial x \partial y}\right|_{P_o} + \delta x \delta z \left.\frac{\partial^2 q_u}{\partial x \partial z}\right|_{P_o}$$

$$+ \delta y \delta z \left.\frac{\partial^2 q_u}{\partial y \partial z}\right|_{P_o} + \delta x \delta y \delta z \left.\frac{\partial^3 q_u}{\partial x \partial y \partial z}\right|_{P_o} + ... + R\left[\delta u^n \left.\frac{\partial^n q_u}{\partial u^n}\right|_{P_o}\right] \tag{5.4}$$

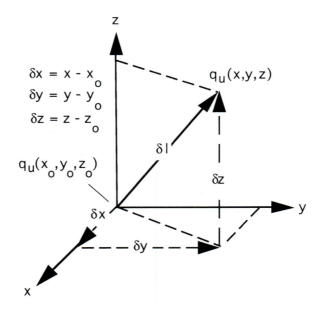

Figure 5.1 The notation for defining the relationship between the component of flux q_u at a point $P_o = (x_o, y_o, z_o)$, given by $q_u(x_o, y_o, z_o)$, and its value at $P = (x,y,z)$, given by $q_u(x,y,z)$.

For our present discussion, we will assume that each flux component is sufficiently smooth and continuous, and each of the offsets δu is sufficiently small, that the higher order derivative terms can be neglected, so that $q_u(x,y,z)$ can be sensibly approximated by the first order expression

$$q_u(x,y,z) \approx q_u(x_o,y_o,z_o) + \delta x \left.\frac{\partial q_u}{\partial x}\right|_{P_o} + \delta y \left.\frac{\partial q_u}{\partial y}\right|_{P_o} + \delta z \left.\frac{\partial q_u}{\partial z}\right|_{P_o} \qquad (5.5)$$

In the following discussion, we will assume that the above relation is exact, and replace the approximate sign by an equal sign. Moreover, we will often simplify the notation in our Taylor's series expansion of a function by using a subscript of "o" to denote those variables evaluated at the local reference point $P_o = (x_o, y_o, z_o)$. Thus the latter expression can be written as

$$q_u = q_{uo} + \delta x \frac{\partial q_u}{\partial x_o} + \delta y \frac{\partial q_u}{\partial y_o} + \delta z \frac{\partial q_u}{\partial z_o} \qquad (5.6)$$

Darcy's Law in Vector Form

For isotropic media, Darcy's law for three dimensional flow at a point can be expressed as the superposition of three vector components

$$\mathbf{q} = -\hat{\mathbf{x}} K \frac{\partial h}{\partial x} - \hat{\mathbf{y}} K \frac{\partial h}{\partial y} - \hat{\mathbf{z}} K \frac{\partial h}{\partial z} \qquad (5.7)$$

where, as before, $\hat{\mathbf{x}}$, $\hat{\mathbf{y}}$ and $\hat{\mathbf{z}}$ are unit vectors in the x, y and z directions respectively.

The latter relation can be expressed in terms of the gradient operator given by

$$\mathbf{grad} = \nabla = \widehat{\mathbf{x}}\frac{\partial}{\partial \mathrm{x}} + \widehat{\mathbf{y}}\frac{\partial}{\partial \mathrm{y}} + \widehat{\mathbf{z}}\frac{\partial}{\partial \mathrm{z}} \qquad (5.8)$$

so that we can relate the specific discharge vector **q** to the hydraulic conductivity K(x,y,z) and the *hydraulic gradient* through

$$\mathbf{q} = - \mathrm{K}\ \mathbf{grad}\ \mathrm{h} \qquad (5.9)$$

or

$$\mathbf{q} = - \mathrm{K}\ \nabla \mathrm{h} \qquad (5.10)$$

where, to simplify notation in the last two expressions, we have suppressed the argument (x,y,z) for each of the spatially dependent parameters.

DIVERGENCE OF FLUX

Continuity of Darcy Flow

The difference between the rate at which a fluid transports mass *out of* a finite volume and the rate at which a fluid transports mass *into* the volume is equal to the rate at which fluid is being generated or lost *within* the volume. Beginning with such a conservation condition

$$\oiint_{\mathrm{Surface}} \mathbf{q} \cdot \mathrm{d}\mathbf{A} = \iiint_{\mathrm{Vol}} [\text{All sources}]\ \mathrm{dV} \qquad (5.11)$$

we apply it to the case of the small rectilinear volume $\Delta \mathrm{V} = \Delta \mathrm{x}\,\Delta \mathrm{y}\,\Delta \mathrm{z}$ in the Figure 5.2, where at the center we locate the origin of a local Cartesian coordinate system, $\mathrm{P_o} = (\mathrm{x_o, y_o, z_o})$.

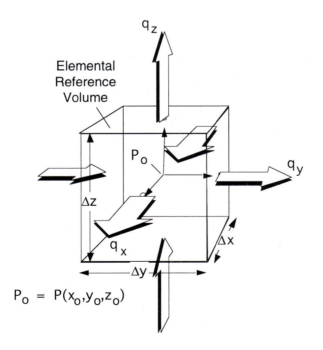

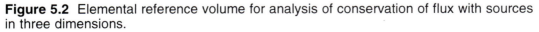

Figure 5.2 Elemental reference volume for analysis of conservation of flux with sources in three dimensions.

Flux Term: Evaluating the LHS of the Conservation Integral

Normal flux through a surface element. Our objective is to evaluate the surface integral on the left hand side of the conservation relation over each of the 6 sides of the rectilinear elemental volume. Without loss of generality, to simplify our expressions, we shift the origin of our local coordinate system to the point P_o, so that $(x_o, y_o, z_o) = (0, 0, 0)$ and, in the following discussion, $\delta x = (x - x_o) = x$, $\delta z = (z - z_o) = z$, and $\delta z = (z - z_o) = z$. A first order Taylor's series representation of $q_x (\Delta x/2, \delta y, \delta z)$, which is the normal component of flux passing through the front surface of the volume, will be

$$q_x = q_{xo} + \frac{\Delta x}{2} \frac{\partial q_x}{\partial x_o} + \delta y \frac{\partial q_x}{\partial y_o} + \delta z \frac{\partial q_x}{\partial z_o} \qquad (5.12)$$

where the subscript "o" denotes that the respective quantity is evaluated at the local origin (x_o, y_o, z_o). The operation is illustrated schematically in Figure 5.3.

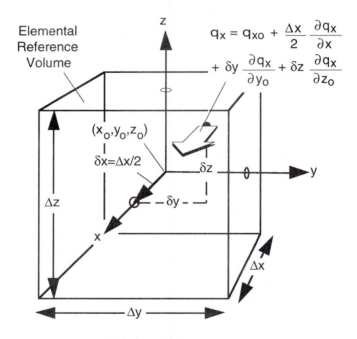

Figure 5.3 The Taylor's series representation of q_x at $\delta x = +\Delta x/2$, over the surface area: $\Delta y \, \Delta z$.

The summation of terms in the Taylor's series for q_x at the surface $\delta x = \Delta x/2$ can be viewed as a construction, or extrapolation, starting with the value of q_{xo}, the value of the x-directed component of flux **q** at the local origin $(x_o, y_o, z_o) = (0, 0, 0)$. The second term represents a step in the x direction a fixed distance $\Delta x / 2$ (fixed by the distance to the mathematical interface of the elemental volume). The third term represents a step in the y direction a variable distance δy, where y is restricted to the surface of the volumes facet. This restricts δy to the range

$$-\Delta y/2 \leq \delta y \leq +\Delta y/2 \qquad (5.13)$$

The fourth term represents a step in the z direction a variable distance δz, where z is restricted to the surface of the volumes facet. This restricts δz to the range

$$-\Delta z/2 \leq \delta z \leq +\Delta z/2 \tag{5.14}$$

A similar expression is obtained for the rear side of the elemental volume, where, in the Taylor's series expansion, $\delta x = -\Delta x/2$. In this case, our expression for $q_x(x,y,z)$ becomes

$$q_x = q_{xo} - \frac{\Delta x}{2}\frac{\partial q_x}{\partial x_o} + \delta y \frac{\partial q_x}{\partial y_o} + \delta z \frac{\partial q_x}{\partial z_o} \tag{5.15}$$

(Note the minus sign before the coefficient of the $\partial q_x/\partial x_o$ term.)

The area element d**A** for each facet of the elemental volume is a vector term directed along the outward normal. For the front face (where $\delta x = +\Delta x/2$), we have

$$d\mathbf{A} = +\hat{\mathbf{x}}\, dA = +\hat{\mathbf{x}}\, dy\, dz \tag{5.16}$$

Whereas, for the rear face (where $\delta x = -\Delta x/2$), we have

$$d\mathbf{A} = -\hat{\mathbf{x}}\, dA = -\hat{\mathbf{x}}\, dy\, dz \tag{5.17}$$

(Note the minus sign in the latter expression, since $\Delta \mathbf{A}$ for the rear facet is directed in the negative x direction.)

We next consider the integral of the normal flux through the front facet (where $\delta x = +\Delta x/2$).

$$\iint_{\substack{\text{FRONT}\\\text{FACE}}} \mathbf{q} \cdot d\mathbf{A} = +\iint_{\substack{\text{FRONT}\\\text{FACE}}} q_x\, dy\, dz = +\iint_{\substack{\text{FRONT}\\\text{FACE}}} \left[q_{xo} + \frac{\Delta x}{2}\frac{\partial q_x}{\partial x_o} + \delta y \frac{\partial q_x}{\partial y_o} + \delta z \frac{\partial q_x}{\partial z_o} \right] dy\, dz \tag{5.18}$$

In the last integral on the right, each of the derivative terms is a constant of the integration, since each represents the fixed value of $\partial q_x/\partial u$ at the local origin (x_o, y_o, z_o). Terms such as δy and δz, however, are variables of integration since δy is a function of y, and δz is a function of z, respectively.

Bringing the various constant terms outside the appropriate integration, results in

$$\iint_{\substack{\text{FRONT}\\\text{FACE}}} \left[q_{xo} + \frac{\Delta x}{2}\frac{\partial q_x}{\partial x_o} + \delta y \frac{\partial q_x}{\partial y_o} + \delta z \frac{\partial q_x}{\partial z_o} \right] dy\, dz = \left[q_{xo} + \frac{\Delta x}{2}\frac{\partial q_x}{\partial x_o} \right] \cdot \iint_{\substack{\text{FRONT}\\\text{FACE}}} dy\, dz +$$

$$+ \left[\frac{\partial q_x}{\partial y_o} \right] \cdot \iint_{\substack{\text{FRONT}\\\text{FACE}}} (\delta y)\, dy\, dz + \left[\frac{\partial q_x}{\partial z_o} \right] \cdot \iint_{\substack{\text{FRONT}\\\text{FACE}}} (\delta z)\, dy\, dz \tag{5.19}$$

We now consider each of the latter three integrals separately. The first integral results in

$$\iint_{\substack{\text{FRONT} \\ \text{FACE}}} dydz = \Delta y \Delta z \tag{5.20}$$

The second integral becomes

$$\iint_{\substack{\text{FRONT} \\ \text{FACE}}} (\delta y)\, dydz = \int_{-\Delta z/2}^{+\Delta z/2} \left[\int_{-\Delta y/2}^{+\Delta y/2} (\delta y)\, dy \right] dz = \int_{-\Delta z/2}^{+\Delta z/2} \left[\left(\frac{(\delta y)^2}{2} \right) \Bigg|_{-\Delta y/2}^{+\Delta y/2} \right] dz \tag{5.21}$$

Since the term in square brackets evaluates to zero, its integral over the range of z is also zero, so that

$$\iint_{\substack{\text{FRONT} \\ \text{FACE}}} (\delta y)\, dydz = 0 \tag{5.22}$$

Similarly, the third integral becomes

$$\iint_{\substack{\text{FRONT} \\ \text{FACE}}} (\delta z)\, dydz = \int_{-\Delta y/2}^{+\Delta y/2} \left[\int_{-\Delta z/2}^{+\Delta z/2} (\delta z)\, dz \right] dy = \int_{-\Delta y/2}^{+\Delta y/2} \left[\left(\frac{(\delta z)^2}{2} \right) \Bigg|_{-\Delta z/2}^{+\Delta z/2} \right] dy \tag{5.23}$$

which results in

$$\iint_{\substack{\text{FRONT} \\ \text{FACE}}} (\delta z)\, dydz = 0 \tag{5.24}$$

Thus the surface integral of the flux through the front facet is given by

$$\iint_{\substack{\text{FRONT} \\ \text{FACE}}} \mathbf{q} \cdot d\mathbf{A} = + \iint_{\substack{\text{FRONT} \\ \text{FACE}}} q_x\, dydz = + \left[q_{xo} + \frac{\Delta x}{2} \frac{\partial q_x}{\partial x_o} \right] \Delta y \Delta z \tag{5.25}$$

It should be readily apparent that the surface integral of the flux through the rear facet is given by

$$\iint_{\substack{\text{REAR} \\ \text{FACE}}} \mathbf{q} \cdot d\mathbf{A} = - \iint_{\substack{\text{REAR} \\ \text{FACE}}} q_x\, dydz = - \left[q_{xo} - \frac{\Delta x}{2} \frac{\partial q_x}{\partial x_o} \right] \Delta y \Delta z \tag{5.26}$$

where the minus sign before the second integral is a result of the outward directed differential area d**A** being in the negative x direction for the rear facet, and the negative sign before $\Delta x/2$ in the third term is a result of projecting the Taylor's series expression to the facet at $\delta x = -\Delta x/2$.

Integration of normal flux through all surface facets. Taking into account the normal flux through all the surface facets of the elemental volume, the left hand side (LHS) of the conservation relation can now be expanded to the form.

$$\oiint_{\text{Surface}} \mathbf{q} \cdot d\mathbf{A} = + \left[q_x\big|_o + \frac{\Delta x}{2} \frac{\partial q_x}{\partial x}\bigg|_o \right] \Delta y \Delta z - \left[q_x\big|_o - \frac{\Delta x}{2} \frac{\partial q_x}{\partial x}\bigg|_o \right] \Delta y \Delta z$$

$$+ \left[q_y\big|_o + \frac{\Delta y}{2} \frac{\partial q_y}{\partial y}\bigg|_o \right] \Delta x \Delta z - \left[q_y\big|_o - \frac{\Delta y}{2} \frac{\partial q_y}{\partial y}\bigg|_o \right] \Delta x \Delta z$$

$$+ \left[q_z\big|_o + \frac{\Delta z}{2} \frac{\partial q_z}{\partial z}\bigg|_o \right] \Delta x \Delta y - \left[q_z\big|_o - \frac{\Delta z}{2} \frac{\partial q_z}{\partial z}\bigg|_o \right] \Delta x \Delta y \qquad (5.27)$$

which simplifies to

$$\text{LHS} = \Delta x \Delta y \Delta z \left[\frac{\partial q_x}{\partial x}\bigg|_o + \frac{\partial q_y}{\partial y}\bigg|_o + \frac{\partial q_z}{\partial z}\bigg|_o \right] \qquad (5.28)$$

Source Term: Evaluating the RHS of the Conservation Integral

The right hand side (RHS) source integral can be expanded according to

$$\text{RHS} = \iiint_{\substack{\text{vol} \\ \text{element}}} \left[w_s\big|_o + \frac{\delta x}{2} \frac{\partial w_s}{\partial x}\bigg|_o + \frac{\delta y}{2} \frac{\partial w_s}{\partial y}\bigg|_o + \frac{\delta z}{2} \frac{\partial w_s}{\partial z}\bigg|_o \right] dx\, dy\, dz \qquad (5.29)$$

which reduces to

$$\text{RHS} = \left[w_s\big|_o \, \Delta x \, \Delta y \, \Delta z \right] \qquad (5.30)$$

The Divergence Condition

Equating the LHS and RHS terms of the conservation relation, we obtain

$$\Delta x \Delta y \Delta z \left[\frac{\partial q_x}{\partial x}\bigg|_o + \frac{\partial q_y}{\partial y}\bigg|_o + \frac{\partial q_z}{\partial z}\bigg|_o \right] = \left[w_s\big|_o \, \Delta x \Delta y \Delta z \right] \qquad (5.31)$$

and cancelling the common elemental volume term $\Delta V = \Delta x \Delta y \Delta z$, we obtain

$$\frac{\partial q_x}{\partial x}\bigg|_o + \frac{\partial q_y}{\partial y}\bigg|_o + \frac{\partial q_z}{\partial z}\bigg|_o = w_s\big|_o \qquad (5.32)$$

- 53 -

Since the position of our local origin (x_0, y_0, z_0) was quite arbitrary, the above differential relation generally holds for any point (x,y,z) in our medium. Thus, we may write what is known as the divergence relation

$$\frac{\partial q_x}{\partial x} + \frac{\partial q_y}{\partial y} + \frac{\partial q_z}{\partial z} = w_s \qquad (5.33)$$

or in terms of the vector operator ∇

$$\nabla \cdot \mathbf{q} = w_s \qquad (5.34)$$

Either of these expressions relates the divergence of the flux at a point to the presence of local sources. If no sources are present, then we have the non-divergence or divergence-free condition

$$\frac{\partial q_x}{\partial x} + \frac{\partial q_y}{\partial y} + \frac{\partial q_z}{\partial z} = 0 \qquad (5.35)$$

or

$$\nabla \cdot \mathbf{q} = 0 \qquad (5.36)$$

CIRCULATION AND CURL OF A FLOW FIELD

Circulation Integral

A measure of the rotational motion of a fluid is given by the following definition of the *circulation integral*:

$$\text{Circulation integral} = \oint \mathbf{q} \cdot d\mathbf{l} \qquad (5.37)$$

which involves the integration of the specific discharge vector **q** along the path of a closed contour generally in three dimensions. To illustrate the importance of (5.37) to groundwater flow, let us construct the integral in the x-y plane for an infinitesimal rectilinear contour centered at the origin of a local coordinate system as shown in Figure 5.4.

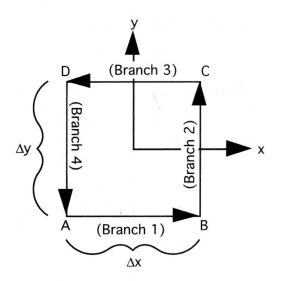

Figure 5.4 The closed contour along which the circulation integral is to be calculated.

Application to an infinitesimal contour. We will assume that the dimensions of the contour are small enough for a first order Taylor's series expansion of each component of **q**, according to (5.6), to be adequate.

Along branch 1 (A-B) we have the line integral:

$$\left[\int_{-\Delta x/2}^{+\Delta x/2} q_x \, dx\right]\Bigg|_{\text{Branch 1}} = \left[q_{xo} - \frac{\Delta y}{2}\frac{\partial q_x}{\partial y}\bigg|_o\right]\int_{-\Delta x/2}^{+\Delta x/2} dx$$

$$= \left[q_{xo} - \frac{\Delta y}{2}\frac{\partial q_x}{\partial y}\bigg|_o\right]\Delta x \tag{5.38}$$

Along branch 2 (B-C):

$$\left[\int_{-\Delta y/2}^{+\Delta y/2} q_y \, dy\right]\Bigg|_{\text{Branch 2}} = \left[q_{yo} + \frac{\Delta x}{2}\frac{\partial q_y}{\partial x}\bigg|_o\right]\Delta y \tag{5.39}$$

Along branch 3 (C-D):

$$\left[\int_{+\Delta x/2}^{-\Delta x/2} q_x \, dx\right]\Bigg|_{\text{Branch 3}} = -\left[q_{xo} + \frac{\Delta y}{2}\frac{\partial q_x}{\partial y}\bigg|_o\right]\Delta x \tag{5.40}$$

Along branch 4 (D-A):

$$\left[\int_{+\Delta y/2}^{-\Delta y/2} q_y \, dy\right]\Bigg|_{\text{Branch 4}} = -\left[q_{yo} - \frac{\Delta x}{2}\frac{\partial q_y}{\partial x}\bigg|_o\right]\Delta y \tag{5.41}$$

Adding up the right hand sides of (5.38) through (5.41) we obtain for the integral (5.37)

$$\oint \mathbf{q}\cdot d\mathbf{l} = \Delta x \, \Delta y \left[\frac{\partial q_y}{\partial x}\bigg|_o - \frac{\partial q_x}{\partial y}\bigg|_o\right] \tag{5.42}$$

It is usual to take the limit when the area of the contour ΔA becomes infinitesimal, or

$$\lim_{\Delta A \to 0}\left[\frac{1}{\Delta A}\oint \mathbf{q}\cdot d\mathbf{l}\right] = \left[\frac{\partial q_y}{\partial x} - \frac{\partial q_x}{\partial y}\right]\bigg|_o \tag{5.43}$$

Curl of the flow field. The right hand side of expression (5.43) is the differential circulation, or *curl*, of the flow field about the z axis at the point P_o. It is often denoted by

$$(\text{curl } \mathbf{q})_z = \frac{\partial q_y}{\partial x} - \frac{\partial q_x}{\partial y} \tag{5.44}$$

where the left hand side is read as "the curl of the the flow field $\mathbf{q}$ about the z axis is equal to ... ".

A simple example. We might apply (5.44) to the simple example of a rigid rotation about the z axis at a radian velocity of

$$\omega = \omega\,\hat{\mathbf{z}} \tag{5.45}$$

as shown in Figure 5.5.

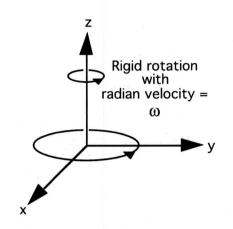

Figure 5.5 Rigid rotation about the z axis at a radian velocity of ω.

We will calculate the curl of the velocity field $\mathbf{v}$ given by

$$\mathbf{v} = \omega \times \mathbf{r} \tag{5.46}$$

where the radius vector is given by

$$\mathbf{r} = x\hat{\mathbf{x}} + y\hat{\mathbf{y}} + z\hat{\mathbf{z}} \tag{5.47}$$

Thus the velocity from the cross product in (5.46) becomes

$$\mathbf{v} = -\omega y\hat{\mathbf{x}} + \omega x\hat{\mathbf{y}} \tag{5.48}$$

Upon substituting the velocity components from (5.48) for their respective terms in the following adaptation of (5.44), we obtain

$$(\text{curl } \mathbf{v})_z = \frac{\partial v_y}{\partial x} - \frac{\partial v_x}{\partial y} \tag{5.49}$$

leading to

$$(\text{curl } \mathbf{v})_z = 2\omega \tag{5.50}$$

In words, (5.50) states that the differential circulation of, or the curl at a point in, a system undergoing a rigid rotation is equal to twice the radian velocity of the rotation.

Application to groundwater flow. Consider flow in the horizontal plane given by the components $\mathbf{q} = (q_x, q_y)$. From Darcy's law $q_x = -K\partial h/\partial x$ and $q_y = -K\partial h/\partial y$. Substituting the right hand side of these relations for the respective components in (5.44), we obtain

$$(\text{curl } \mathbf{q})_z = -K\frac{\partial}{\partial x}\left(\frac{\partial h}{\partial y}\right) + K\frac{\partial}{\partial y}\left(\frac{\partial h}{\partial x}\right) \equiv 0 \qquad (5.51)$$

In other words, *the curl of the vector flow field in a plane is zero.* This is a consequence of the fact that, according to Darcy's law, the specific discharge $\mathbf{q}$ can be derived from the gradient of the potential field h.

The curl in three dimensions. In general, the curl of a flow field can have components along any or all of the three cartesian axes, so that the curl becomes a vector quantity that can be expressed in the form

$$\text{curl } \mathbf{q} = \left(\frac{\partial q_z}{\partial y} - \frac{\partial q_y}{\partial z}\right)\widehat{\mathbf{x}} + \left(\frac{\partial q_x}{\partial z} - \frac{\partial q_z}{\partial x}\right)\widehat{\mathbf{y}} + \left(\frac{\partial q_y}{\partial x} - \frac{\partial q_x}{\partial y}\right)\widehat{\mathbf{z}} \qquad (5.52)$$

An alternative notation for the curl is

$$\nabla \times \mathbf{q} = \text{curl } \mathbf{q} \qquad (5.53)$$

Specific discharge as a curl-free vector field. Inspection of the various differential terms in (5.52) should indicate that the conclusion drawn from (5.51) pertains to each of the components of the vector curl.

We conclude, therefore, that in general, because the specific discharge $\mathbf{q}$ for groundwater flow can be derived from the gradient of a potential field h(x,y,z), the curl of $\mathbf{q}$ is always zero, or

$$\nabla \times \mathbf{q} = \text{curl } \mathbf{q} = 0 \qquad (5.54)$$

This is a fundamental constraint on any mathematical expressions representing groundwater flow.

THE FUNDAMENTAL EQUATION OF GROUNDWATER FLOW

Begin with *Darcy's Law*

$$\mathbf{q} = -K\nabla h \qquad (5.55)$$

where K is the hydraulic conductivity and h is the hydraulic head. Next, recall the *divergence relation*

$$\nabla \cdot \mathbf{q} = w_s \qquad (5.56)$$

or

$$\partial q_x/\partial x + \partial q_y/\partial y + \partial q_z/\partial z = w_s \qquad (5.57)$$

where

$$\mathbf{q} = (q_x, q_y, q_z) \qquad (5.58)$$

is the specific discharge. Note that w_s represents all classes of volumetric sources.

Substituting *Darcy's law* (5.55) into the left hand side of the *divergence relation* (5.56) we obtain the following fundamental equation for groundwater flow

$$\nabla \cdot [- K\nabla h] = w_s \qquad (5.59)$$

which we will often refer to as the "flow equation".

In the case of a homogeneous, isotropic medium, K is constant, so that the latter expression reduces to a form of the flow equation that is often referred to as **Poisson's equation**:

$$\nabla^2 h = - (w_s / K) \qquad (5.60)$$

or

$$\frac{\partial^2 h}{\partial x^2} + \frac{\partial^2 h}{\partial y^2} + \frac{\partial^2 h}{\partial z^2} = - \frac{w_s}{K} \qquad (5.61)$$

In the absence of local sources ($w_s = 0$), the latter reduces the flow equation to the following form of **Laplace's equation**:

$$\nabla^2 h = 0 \qquad (5.62)$$

or

$$\frac{\partial^2 h}{\partial x^2} + \frac{\partial^2 h}{\partial y^2} + \frac{\partial^2 h}{\partial z^2} = 0 \qquad (5.63)$$

Generic Source Terms in the Flow Equation

In the fundamental relation for groundwater flow:

$$\nabla \cdot [- K\nabla h] = w_s \qquad (5.64)$$

the term w_s is a generic term representing all classes of volumetric sources. In general, w_s may be steady state (not a function of time), or transient (periodic or aperiodic). We thus expand the flow equation to the form:

$$
\begin{aligned}
\partial[K\partial h/\partial x]/\partial x + \partial[K\partial h/\partial y]/\partial y &+ \partial[K\partial h/\partial z]/\partial z = \\
&- w_{steady\text{-}state}(x,y,z) - w_{storage}(x,y,z,t) \\
&- w_{infiltration}(x,y,z,t) - \bullet \bullet \bullet \qquad (5.65)
\end{aligned}
$$

which explicitly identifies some (but not all) of the possible source terms. Note that this relation may be applied to inhomogeneous isotropic media (also known as nonuniform or heterogeneous isotropic media).

PART II. STEADY-STATE FLOW

Chapter 6. Two Dimensional Steady-State Flow

FLOW FUNCTIONS AND FLOW NETS

The study of what are called "stream functions" or "flow functions" is a branch of mathematical analysis that dates back to the 19th century, and is used by physicists, engineers, and hydrologists to parameterize the flow of mass and energy in two dimensions as shown in Figure 6.1. A rich literature exists in the fields of not only fluid flow and hydraulics, but electricity and magnetism, heat flow, and aerodynamics, as well. In the present text, we will apply some of these concepts to groundwater flow.

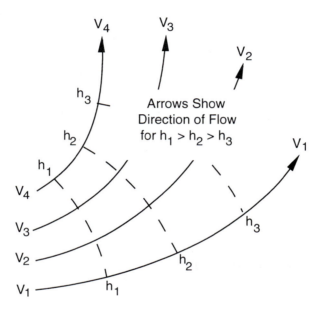

Figure 6.1 A simple flow-net where the solid curved lines show the direction of fluid flow orthogonal to lines of constant hydraulic head for $h_1 > h_2 > h_3$. The corresponding flow-lines are labelled V_1, V_2, V_3 and V_4, respectively.

We assume two dimensional, steady-state flow under source-free conditions. The geometry of the subsurface flow can be either in the vertical plane, such as a hydrogeological cross-section, or in a horizontal plane, such as in an aquifer of uniform thickness. (In principle, flow can also be in a generalized curvilinear sheet in three dimensions, but this case is not considered in this text.)

In the following, we will assume for specificity that flow is in the horizontal or x-y plane, although it is a trivial matter to redefine the coordinates for the vertical x-z or y-z plane.

Darcy's law in general form

$$\mathbf{q} = - K \nabla h \tag{6.1}$$

reduces, in two dimensions, to

$$\mathbf{q} = -K \frac{\partial h}{\partial x} \hat{\mathbf{x}} \ -K \frac{\partial h}{\partial y} \hat{\mathbf{y}} \tag{6.2}$$

where the specific discharge **q** has components

$$\mathbf{q} = q_x \hat{\mathbf{x}} + q_y \hat{\mathbf{y}}$$

(6.3)

In addition, the conservation of flux at a point, for source free conditions, leads to the following form of zero divergence

$$\nabla \cdot \mathbf{q} = 0$$

(6.4)

which, in two dimensions, reduces to the form

$$\frac{\partial q_x}{\partial x} + \frac{\partial q_y}{\partial y} = 0$$

(6.5)

We shall see that these conditions lead to an orthogonal relation between flow-lines (or "stream-lines" in some of the more general literature) and h-lines that can be used to construct "flow-nets". Observing the rules of subsurface flow, such flow-nets may be constructed analytically as closed-form mathematical expressions, or by hand, or by computer codes that solve the appropriate flow equations numerically. By whatever means one uses to construct such relations, the following assumptions are generally made:

- Steady-state, source-free, saturated flow in two dimension
- An isotropic, homogeneous medium
- Boundary conditions are completely prescribed, and involve a set of one or more of the following types:

    ⇒ Constant head boundary

    ⇒ Water-table (free surface) boundary

    ⇒ Variable in space, but fixed in time, head boundary

    ⇒ No-flow boundary

    ⇒ Fixed-flow boundary

Analytical Basis of Flow Nets

A little reflection will indicate that, according to the above two-dimensional divergence condition (6.5), the flux components q_x and q_y, have a rather interesting property. If the two-dimensional divergence is, in fact, zero, then each of the flux components can be derived from a common function that we will denote by V(x,y), such that

$$q_x = \frac{\partial V}{\partial y}$$

(6.6)

and

$$q_y = -\frac{\partial V}{\partial x}$$

(6.7)

Clearly substituting (6.6) and (6.7) into (6.5) leads to

$$\frac{\partial}{\partial x}\left[\frac{\partial V}{\partial y}\right] - \frac{\partial}{\partial y}\left[\frac{\partial V}{\partial x}\right] = 0$$

(6.8)

which verifies our premise. This special function V(x,y) is known as a "flow function" or, synonymously, as a "stream function" in mathematical physics (e.g. Morse and Feshbach, 1953, p. 154; or Landau and Lifshitz, 1959, p. 20). More advanced discussions are often couched in terms of functions of a complex variable (e.g. Bear, 1988), however, for purposes of the present text, the basic principles are readily illustrated using only real functions of the form V(x,y).

Note on terminology: In keeping with much of the current hydrogeological literature dealing with the terms flow-lines and h-lines, we will refer to the function V(x,y) as a "flow function", from which, as we will see below, one determines "flow-lines", or vice-versa.

Lest it escape the reader's attention, note the result that the analytical properties of V(x,y) embodied in (6.6) and (6.7) lead to the result that

q_x, the flux in the x direction, depends on the directional gradient of V in the *y direction*,
q_y, the flux in the y direction, depends on the directional gradient of V in the *x direction*.

To obtain more insight into the interrelationship of these parameters, rewrite (6.6) and (6.7) in differential form along a line of constant x

$$q_x \, dy = dV \tag{6.9a}$$

or along a line of constant y

$$q_y \, dx = - \, dV \tag{6.9b}$$

where of course the dV on the right hand side of each of the latter expressions may be different. Recalling that a line element in 2-D is analogous to a cross-sectional area in 3-D, expressions (6.9a) and (6.9b) emphasize the concept that the flux (e.g. q_x in (6.9a)) flowing normal to a finite line element (e.g. dy) is equal to the incremental change in V, which is related, in turn, to the incremental change in the total two-dimensional discharge Q_{2D}, by

$$dV = dQ_{2D} \tag{6.10a}$$

Thus dV is the incremental flow across the y axis (given by (6.9a)), or the incremental flow across the x axis (given by (6.9b)). If the incremental path has both a dx and a dy component, the total incremental flow is given by

$$dV = dQ_{2D} = q_x \, dy - q_y \, dx. \tag{6.10b}$$

The minus sign before the q_y dx term on the right is needed to assure that *positive* flow is associated with flow directed *outward* crossing a closed contour in the x-y plane. By this convention, the first term, q_x dy, is a positive flow when both q_x and dy are *positively directed*; however the second term is positive when either q_y or dx is *negatively directed*.

With reference to Figure 6.2, the total two-dimensional flow crossing a contour (actually a surface of unit thickness or depth in 3-D) drawn from point A to point B is given by

$$Q_{2D} = \int_A^B \mathbf{q} \cdot \hat{\mathbf{n}} \, dl \tag{6.11}$$

where the unit normal vector is given by $\hat{\mathbf{n}} = d\mathbf{l} \times \hat{\mathbf{z}} \, / \, |\, d\mathbf{l} \,|$.

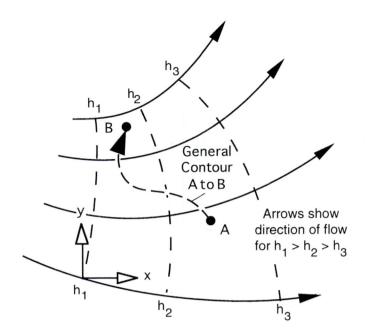

Figure 6.2 Total two-dimensional flow crossing the contour A-B. This 2-D contour is the trace in the x-y plane of a 3-D surface of unit depth.

Note on units: The units of the specific discharge q in two-dimensional flow are still in terms of the volume of water crossing a unit area per unit time. [The SI units are $m^3 s^{-1} m^{-2}$ or $m^1 s^{-1}$; q has general dimensions of $L^3 L^{-2} T^{-1} = L T^{-1}$, or length per unit time.]. However, when describing the *total discharge* Q crossing a boundary in 2-D, one has to be careful. In 2-D, a *line* element becomes a surrogate to a *cross-sectional area* element in 3-D. To make 2-D and 3-D representations consistent, one assumes that any *line element* in 2-D, having a length ΔL, also has *a unit thickness* (such as one meter for SI units). In other words, if a flux or specific discharge, for example q_x, crosses normal to a line element of length $\Delta L = \Delta y$, the integrated flux is $q_x \Delta y$, which has units of $(m s^{-1} m)$ or $m^2 s^{-1}$; or in terms of the total 2D discharge Q_{2D}, the units correspond to the total volume (m^3) per unit time (s^{-1}) per unit thickness (m^{-1}). For the latter, we have $m^3 s^{-1} m^{-1}$, which again reduces to $m^2 s^{-1}$. Thus, the SI units for Q_{2D} will be $m^2 s^{-1}$, and Q_{2D} will have general dimensions of $[L^2 T^{-1}]$. For any 2-D problem, we will assume that if one wants the true 3-D flow volume, then they will simply multiply the 2-D discharge Q_{2D} by an appropriate invariant "third-dimension". For strictly horizontal flow in the x-y plane, this might be the thickness of the aquifer. In the case of flow in the vertical plane (either the x-z or y-z plane), this might involve multiplying by some appropriate dimension along the strike of the feature. In regional flow, it may be the length of a mountain ridge delivering water to a watershed. Or, if one were analyzing 2-D seepage through an earthen dam, they might want to estimate how much water they are losing from an impounded reservoir, so an appropriate "third-dimension" might be the length of the dam perpendicular to flow. In short, in 2-D the units of q continue to be $m s^{-1}$ [dimensions: $L T^{-1}$], but the units of the total discharge Q_{2D} are $m^2 s^{-1}$ [dimensions: $L^2 T^{-1}$].

Recall that the components of d**l** are given in terms of the following products of direction cosines and the differential elements dx and dy:

$$dx = dl \cos \alpha \qquad (6.12)$$
$$dy = dl \cos \beta \qquad (6.13)$$

Thus (6.11) can be written as

$$Q_{2D} = \int_A^B \left[q_x dy - q_y dx \right]$$ (6.14)

Suppose we divide the general line integral A-B into the two specific branches shown in Figure 6.3: first a branch from A to A', then a branch from A' to B.

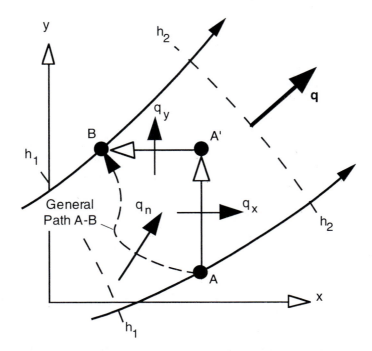

Figure 6.3 Reducing the general contour path A-B to two branches: A-A' and A'-B. Clearly q_x is the component of flow **q** normal to A-A', and q_y is the component of flow normal to A'-B.

The first goes from A to A' along the y-axis, leading to

$$Q_{2D}^{A\text{-}A'} = \int_A^{A'} q_x dy = \int_A^{A'} dV$$ (6.15)

The next branch goes from A' to B along the x-axis, leading to

$$Q_{2D}^{A'\text{-}B} = -\int_{A'}^B q_y dx = \int_{A'}^B dV$$ (6.16)

We can see that, for two-dimensional flow, the total flux crossing a contour between A and B is independent of path by the following construction. First, we recall that the flux crossing a closed contour in 2-D is analogous to the flux crossing a surface in 3-D (see Figure 6.4).

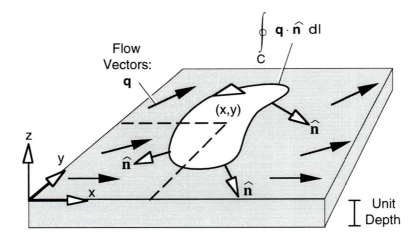

Figure 6.4 Analogy between flux crossing a closed 2-D contour C in the x-y plane, and flux crossing a closed surface of unit depth in 3-D.

Recall that the line integral of flux or specific discharge crossing each elemental length (normal $\mathbf{q}$ times path length) is the rate of total volume flow per unit time per unit depth crossing the entire path, and that, in 2-D, flow is restricted to the x-y plane, so that the integral of normal flux escaping through the top and bottom of a true 3-D surface as shown in Figure 6.4 is implicitly zero. Thus a conservation condition in 2-D, for a source free region, can be written in integral form as

$$\oint_c \mathbf{q} \cdot \hat{\mathbf{n}} \, dl = 0 \tag{6.17}$$

where the unit normal vector is given by $\hat{\mathbf{n}} = d\mathbf{l} \times \hat{\mathbf{z}} / |\, d\mathbf{l}|$.

Consider, now, the conservation of flux through the *closed* contour C shown in Figure 6.5, that extends from A to A' to B, then back to A.

The integral along C becomes

$$\oint_c \mathbf{q} \cdot \hat{\mathbf{n}} \, dl = 0 = \int_A^{A'} q_x \, dy - \int_{A'}^{B} q_y \, dx + \int_B^{A} \mathbf{q} \cdot \hat{\mathbf{n}} \, dl \tag{6.18}$$

Upon rearranging (6.18), recalling that switching the limits of integration on the last term introduces a negative sign on the integral, we obtain

$$Q_{2D} = \int_A^{B} \mathbf{q} \cdot \hat{\mathbf{n}} \, dl = \int_A^{A'} q_x \, dy - \int_{A'}^{B} q_y \, dx \tag{6.19}$$

which, with a little thought, is seen as essentially a statement that the total flux passing between two points A and B is independent of the path of integration. (Note that here and in the following discussion, Q_{2D} will be understood to represent the total two-dimensional flux or discharge passing between the two points A and B.)

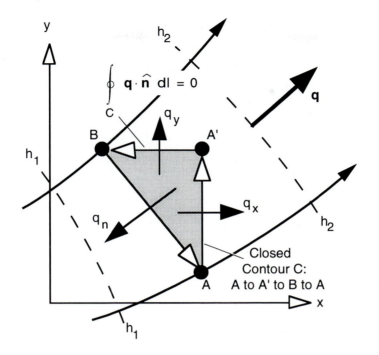

Figure 6.5 Conservation of flux through the *closed* contour C extending from A to A' to B, then back to A. Contour C is the trace in the x-y plane of a 3-D surface of unit depth.

To understand the implications of (6.19), assume the contour that we are considering is of differential dimensions (see Figure 6.6).

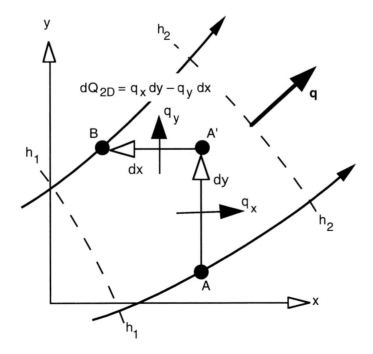

Figure 6.6 A contour, A to A' to B, of differential dimensions.

In this case

$$dQ_{2D} = dQ_{2D}^{A\text{-}A'} + dQ_{2D}^{A'\text{-}B} \tag{6.20}$$

and

$$dQ_{2D} = q_x \, dy - q_y \, dx \tag{6.21}$$

which is the total 2D discharge crossing the differential line element from A to B. Clearly, (6.19) is simply the total integral of (6.21) when the distance from A to B becomes finite.

Using (6.15) and (6.16) to rewrite (6.19) in terms of the integral over the total differential dV, we obtain

$$Q_{2D} = Q_{2D}^{A\text{-}A'} + Q_{2D}^{A'\text{-}B} = \int_{A}^{A'} dV + \int_{A'}^{B} dV \tag{6.22}$$

which reduces to

$$Q_{2D} = \int_{A}^{B} dV \tag{6.23}$$

Integrating the total differential from A to B

$$\int_{A}^{B} dV = V_B - V_A \tag{6.24}$$

we see that its value is independent of path, and depends only on the values of V at its end points. This is typical of all scalar potential functions, from temperature to voltage.

Combining (6.23) and (6.24) in the form

$$Q_{2D} = V_B - V_A \tag{6.25}$$

leads to the conclusion that the total flow between two points A and B is given by the difference in the flow-function V(x,y) at those two points.

In other words, either (6.6) and (6.7) in differential form, or (6.25) in integral form, can be used to determine the flux or specific discharge, providing one is given the flow function V(x,y).

Alternatively, reversing the order of (6.25), we obtain

$$V_B - V_A = Q_{2D} \tag{6.26}$$

which underscores the conclusion that the difference in V(x,y) between two points is given by the total discharge Q_{2D} flowing between them.

Determining Flow Functions Analytically

Relation (6.26), along with (6.14), points the way to determining V(x,y) over a particular region of space, given the vector field of specific discharge $\mathbf{q}$(x,y) over the same region. Starting at a specific point A, where we assume we "know" V_A (or can arbitrarily set $V_A = 0$), one can use the integral

$$V_B = V_A + \int_A^B \left[q_x dy - q_y dx \right] \tag{6.27}$$

to determine the value of the flow function V(x,y) at a point B. Since this point is arbitrary within the region that $\mathbf{q} = (q_x, q_y)$ is defined, we can thus set V(x,y) = V_B, and write

$$V(x,y) = V_A + \int_{(x_A, y_A)}^{(x,y)} \left[q_x dy - q_y dx \right] \tag{6.28}$$

where V_A is the assigned value of the flow function at point A or (x_A, y_A). Often A is specified as the origin, and V_A is arbitrarily set equal to 0.

An alternative form to (6.28) is to use an indefinite integral

$$V(x,y) = \int \left[q_x dy - q_y dx \right] \tag{6.29}$$

The resulting constant of integration implied by (6.29) may be understood to be the value of the flow function at some assigned reference point, such as the origin (0,0).

Since $\mathbf{q}$(x,y) can, in turn, be derived from the hydraulic head h(x,y) using Darcy's law (6.2), one may expand (6.29) to the form

$$V(x,y) = -K \int \left[\frac{\partial h}{\partial x} dy - \frac{\partial h}{\partial y} dx \right] \tag{6.30}$$

where K is the hydraulic conductivity of the medium, which we assume here is homogeneous.

Flow-Lines

The last section developed the concept that V(x,y) is a scalar function of x and y that can be used to determine the specific discharge vector $\mathbf{q} = (q_x, q_y)$ at any point in space. From (6.6) and (6.7), it is clear that directional gradients in V(x,y) are responsible for generating a corresponding flow. Consider now the special class of points, x and y, that trace out the trajectory of a curve

$$V(x,y) = \text{constant} \tag{6.31}$$

or, to be specific, say

$$V(x,y) = V_1 \tag{6.32}$$

Since V is constant, the total differential

$$dV = 0 \qquad (6.33)$$

along the path given by (6.32). Recalling (6.10b)

$$dV = dQ_{2D} = q_x \, dy - q_y \, dx. \qquad (6.10b)$$

upon substituting (6.33), we obtain

$$dQ_{2D} = 0 \qquad (6.34)$$

So that $V(x,y) = V_1$ represents a curve in space across which there is no flux. This curve is termed a "flow-line" corresponding to the specified value of V_1. Setting $V(x,y)$ equal to another constant will yield another trajectory of points representing another flow-line across which there is no flux.

When flow-lines are plotted for a specific situation, usually arrows are attached to the end of the lines, or in some cases at various points along each line, to show the direction of fluid flow. This would be the direction of the negative gradient of the hydraulic head. One should keep in mind, however, that $V(x,y)$ is strictly a *scalar* function of space.

Flux-Tubes

If we mathematically construct another flow-line by assigning $V(x,y)$ a constant value V_2 which is different from V_1 by a small but finite value ΔV, such that

$$V_2 = V_1 + \Delta V \qquad (6.35)$$

then we have defined two curves in space that, in effect, bound the fluid flow. Each of the curves, $V(x,y) = V_1$ and $V(x,y) = V_2$, is a flow-line, and the region between the two closely spaced flow-lines is termed a "flux tube", since it is a mathematical boundary whose walls, in a sense, guide the transport of fluid through space (see Figure 6.7). We could, in fact, introduce thin, impermeable membranes along the walls of such a flux-tube and, neglecting viscous losses along the membrane walls of the tube, there would be absolutely no difference in the overall flow regime of the fluid. We could take this a step further and, providing we keep the hydraulic head conditions the same along the walls and on the two ends of the flux-tube, we could take away all the material outside the flux-tube, and flow conditions inside the tube would remain precisely the same.

One can mathematically construct a family of n flux tubes bounded by flux-lines such that

$$V_i = V_o + i \, \Delta V \qquad (6.36)$$

for i = 0 to n, where ΔV is assigned a convenient constant value. Thus the discharge or total flux between each flow-line or within each flux-tube is constant in terms of volume per unit time per unit depth. If one is given the distribution, or a map, of the flux-lines and the value of ΔV for a particular situation, they can then find the total discharge between two arbitrary points by simply adding up the number of flux-tubes between the points — no formal integration is necessary. As an example, for n flux-tubes between the point A and B, the total flux is given by

$$Q_{2D} = n \, \Delta V \qquad (6.37)$$

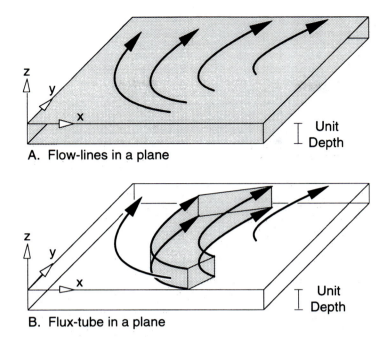

Figure 6.7 Showing how flux-lines in a plane (Panel A) can be used to prescribe a flux-tube (Panel B).

h-Lines

We've discussed in earlier chapters the concept of hydraulic head h, and the fact that for an isotropic medium, a consequence of Darcy's law is that flow is in the direction of the greatest spatial rate of decrease of h(x,y); this is, after all, the physical implication of the negative gradient. For steady-state flow, the field of hydraulic head h(x,y) is a potential field, and its negative gradient leads to a current flow, precisely analogous to the relation between electric voltage and current. Just as we speak of equipotentials in electricity, we can use constant values of hydraulic head

$$h(x,y) = \text{constant} \qquad (6.38)$$

or more specifically

$$h(x,y) = h_1 \qquad (6.39)$$

to represent a trajectory of points in x and y space that is an equipotential of hydraulic head, often termed an "*h-line*" or simply an *equipotential* line (or surface).

Recall that the total differential of h(x,y) at a general point in space in an arbitrary differential direction

$$d\mathbf{l} = dx\hat{\mathbf{x}} + dy\hat{\mathbf{y}} \qquad (6.40)$$

is given by

$$dh = \frac{\partial h}{\partial x} dx + \frac{\partial h}{\partial y} dy \qquad (6.41)$$

or

$$dh = \nabla h \cdot d\mathbf{l} \qquad (6.42)$$

If the differential offset element d**l** is taken along an h-line, then since h = constant, we have

$$dh = 0 \qquad (6.43)$$

and that

$$\nabla h \cdot d\mathbf{l} = 0 \qquad (6.44)$$

The latter expression implies that a step along an h-line is orthogonal to the gradient of h. This is clearly seen from the definition of the dot product (or the projection of d**l** on to ∇ h):

$$\nabla h \cdot d\mathbf{l} = \left| \nabla h \right| \left| d\mathbf{l} \right| \cos\theta \qquad (6.45)$$

For this product to be zero, for the non-trivial case when neither the gradient nor the offset is zero, we must have $\theta = 90^\circ$. The consequence of this is that, since flow is in the direction of the negative gradient, and there is no gradient along an h-line, then there *is no flow along an h-line*.

On the other hand, one may see that this product is maximum when θ is 0°, which in turn implies that the elemental offset d**l** must be in the direction of the vector gradient ∇ h. Thus, the maximum rate of increase of h is in the direction of the principal unit vector

$$\hat{\mathbf{u}}_p = \frac{d\mathbf{l}_p}{\left| d\mathbf{l}_p \right|} = \frac{\hat{\mathbf{x}} \, \partial h/\partial x + \hat{\mathbf{y}} \, \partial h/\partial y}{\left(\left(\partial h/\partial x \right)^2 + \left(\partial h/\partial y \right)^2 \right)^{1/2}} \qquad (6.46)$$

along which the maximum flow is oppositely directed.

Construction of Flow Nets

Flow nets are the superposition of flow-lines and h-lines for a given flow condition in order to visually portray the path and intensity of fluid flow in two dimensions. While such a presentation can be synthesized in various ways, and at various scales of compression or enhancement, over the years hydrogeologists have developed certain rules or protocols for what we will term here a "classical" flow-net construction — this is to develop a plan view map, or vertical cross-section, showing a pattern of flow-lines and orthogonal intersecting h-lines according to the following procedure.

- The scales for the two dimensions, x and y, or horizontal and vertical, must be the same. (That is to say, no vertical exaggeration is permitted.)

- A set of consistent units is selected for the hydraulic conductivity, the flow-lines and the h-lines. In the SI system, the units would be meters and seconds, so that $[K] = m \, s^{-1}$, $[V] = m^2 \, s^{-1}$, $[h] = m$.

- The set of flow-lines intersect the set of h-lines in such a way that a pattern of quasi-rectilinear, equidimensional, quadrilateral cells ("quasi-squares") is formed. (Note: The choice of quasi-squares is often one of aesthetics, but it may also have a utilitarian purpose for interpretation as discussed below. In some applications, however, one may favor a higher density of flow-lines over h-lines, or vice versa, for specific technical reasons.)

- The orthogonal contours are spaced according to the following conventions:
  ⇒ Flow-lines are plotted such that the same flux flows between adjacent surfaces;
  ⇒ h-lines are plotted such that adjacent contours differ by the same increment in hydraulic head.

Quantitative Interpretation of Flow-Nets

Providing the above conventions are followed, and if the flow-lines are actually calibrated from an analytical or numerical calculation, one may read the total discharge Q_{2D} directly from the flow-net through

$$Q_{2D} = n_v \, \Delta V \qquad (6.47)$$

where n_v is the number of flux-tubes between the two points and ΔV is the interval between the flow-line contours. (Note: Recall that the units of ΔV will be in total volume per unit time for a unit depth or unit thickness; for example, $[\Delta V] = m^3 \, s^{-1} \, m^{-1} = m^2 \, s^{-1}$).

An alternative procedure is to employ an adaptation of Darcy's law to analyze the relation among Q_{2D}, flow-lines, h-lines, and head loss. To do so, one should have structured their flow net into the "quasi-square" format for the "cells" that are prescribed between adjacent pairs of h-lines and quasi-orthogonal flow-lines. This assures that the distance scales are the same for the V-lines and h-lines. Thus, the following expression allows one to relate total flow Q_{2D} to head loss directly from inspecting the flow-net:

$$Q_{2D} = K \, n_v \, \Delta h_{loss} \, / \, n_h \qquad (6.48)$$

where

  K is the hydraulic conductivity,
  Q_{2D} is the total discharge between n_v flux tubes (in units of volume per time per unit depth or width of the 2-D flow),
  Δh_{loss} is the *total* head loss across n_h h "cells" (not to be confused with the fixed increment between *adjacent* h-lines).

This relation may be used to calculate partial flows, or may be integrated (summed) to find total flow, for example, the total leakage from a reservoir through an earthen dam, etc.

Note: The above protocols are usually invoked in order to portray the most quantitative information in a "classical" flow-net. For specific applications, a hydrogeologist may find other formats more suitable; but these would not be the "classical" form.

Analytical Example of a Flow-Net

A form for V(x,y). To add substance to our discussion, we now consider from the start the generation of a set of relations representing a specific flow regime in two dimensions. To do so, we will simply pull a functional form for $V(x,y)$ "out-of-the-air", and evaluate its properties; first to see if it is a proper functional form, then to analyze its associated flow regime. In going through such an exercise, it is often useful to start, not with the most complicated case one could imagine, but with the simplest case that might have even a remote chance for showing some interesting physics. We thus assume that $V(x,y)$ can be represented by a simple polynomial of the form

$$V(x,y) = A + Bx + Cy + Dxy \qquad (6.49)$$

Determining q from V(x,y). That this is a proper form can be seen from applying (6.6), (6.7) and (6.8), along with assuring that any derived form for **q** is curl free. First we see that we can use (6.6) and (6.7) to derive the following expressions for q_x and q_y:

$$q_x = \frac{\partial V}{\partial y} = C + Dx \qquad (6.50)$$

$$q_y = -\frac{\partial V}{\partial x} = -B - Dy \qquad (6.51)$$

These expressions indicate for the flow function (6.49) that the specific discharge, or flux, varies such that q_x changes linearly in the x direction and q_y changes linearly in the y direction. Which component increases and which decreases depends on assigning specific values to the constants.

While q_x and q_y in (6.50) and (6.51), respectively, appear "reasonable", are they in fact physically acceptable? In other words, do they conform to the appropriate fundamental laws of physics?

Curl-free condition. One condition to be met is whether the forms for q_x and q_y derived from V(x,y) satisfy the requirement that the specific discharge be curl free. For flow in the x-y plane, this requires that

$$(\text{curl } \mathbf{q})_z = \frac{\partial q_y}{\partial x} - \frac{\partial q_x}{\partial y} \equiv 0 \qquad (6.52)$$

which is clearly met by substituting the right hand sides of (6.50) and (6.51) for the appropriate terms in (6.52), so that

$$(\text{curl } \mathbf{q})_z = \left[\frac{\partial}{\partial x}(-B - Dy) \right] - \left[\frac{\partial}{\partial y}(C + Dx) \right] \equiv 0 \qquad (6.53)$$

Divergence-free condition. Another condition is that for the case of steady-state flow under source-free conditions, we expect that net flux through any reference volume is conserved. As discussed earlier in this chapter, this requires that q_x and q_y satisfy the divergence condition:

$$\frac{\partial q_x}{\partial x} + \frac{\partial q_y}{\partial y} = 0 \qquad (6.54)$$

Do (6.50) and (6.51), when substituted into (6.54), satisfy the latter condition? By making the substitution, we obtain

$$\frac{\partial}{\partial x}\left[\frac{\partial V}{\partial y}\right] - \frac{\partial}{\partial y}\left[\frac{\partial V}{\partial x}\right] = \frac{\partial}{\partial x}[C + Dx] - \frac{\partial}{\partial y}[B + Dy] = D - D = 0 \qquad (6.55)$$

confirming that condition (6.8)

$$\frac{\partial}{\partial x}\left[\frac{\partial V}{\partial y}\right] - \frac{\partial}{\partial y}\left[\frac{\partial V}{\partial x}\right] = 0 \qquad (6.8)$$

on V(x,y) is satisfied. We thus conclude that V(x,y) given by (6.49) is a proper form, and that the associated vector components of the specific discharge vector **q** are given by (6.50) and (6.51), where the constants B, C and D are arbitrary, and can be adjusted to correspond to the particular situation to be simulated.

Determining h(x,y) from V(x,y). We next want to determine a form for the hydraulic head $h(x,y)$ that is consistent with our assumed flow function V(x,y). Given the form $\tilde{V}(x,y)$ in (6.49), from which we determined the flow components (6.50) and (6.51), we can use these latter two expressions, along with Darcy's law, to determine the following partial derivatives of the hydraulic head.

We can rearrange the expression for q_x given by

$$q_x = -K \frac{\partial h}{\partial x} = C + Dx \qquad (6.50)$$

to the form

$$\frac{\partial h}{\partial x} = -\frac{1}{K}(C + Dx) \qquad (6.56)$$

And we can rearrange the expression for q_y given by

$$q_y = -K \frac{\partial h}{\partial y} = -B - Dy \qquad (6.51)$$

to the form

$$\frac{\partial h}{\partial y} = \frac{1}{K}(B + Dy) \qquad (6.57)$$

The two expressions, (6.55) and (6.57), represent differential forms for h based on the particular flow function V(x,y) we have selected. We next want to integrate these *differential* expressions to obtain a closed form for $h(x,y)$ that is consistent with the initial flow function (6.49). To do so, we recall some general relations from the calculus of two variables.

The total differential of a general function F(x,y) is given by

$$dF = \frac{\partial F}{\partial x} dx + \frac{\partial F}{\partial y} dy \qquad (6.58)$$

If we use the notation that P(x,y) is given by

$$P(x,y) = \frac{\partial F}{\partial x} \qquad (6.59)$$

and Q(x,y) is given by

$$Q(x,y) = \frac{\partial F}{\partial y} \qquad (6.60)$$

the total differential can be written in the form

$$dF = P(x,y) \, dx + Q(x,y) \, dy \qquad (6.61)$$

and its definite integral becomes

$$F(x,y) = \int_A^B dF = \int_A^B [\, P(x,y)\, dx + Q(x,y)\, dy\,] \qquad (6.62)$$

Expression (6.62) is a fundamental relation obtained in most advanced calculus texts. We now apply it to an expression for the total differential of h(x,y) given by

$$dh = \frac{\partial h}{\partial x}\, dx + \frac{\partial h}{\partial y}\, dy \qquad (6.63)$$

where we substitute

$$P(x,y) = \frac{\partial h}{\partial x} \qquad (6.64)$$

and

$$Q(x,y) = \frac{\partial h}{\partial y} \qquad (6.65)$$

into

$$dh = P(x,y)\, dx + Q(x,y)\, dy \qquad (6.66)$$

Note that (6.64) is the left hand side of (6.56), and (6.65) is the left hand side of (6.57). In analogy to (6.62), the integral of (6.66) becomes

$$h(x,y) = \int_A^B dh = \int_A^B [\, P(x,y)\, dx + Q(x,y)\, dy\,] \qquad (6.67)$$

which can be broken down into the two indefinite integrals

$$\int P(x,y)\, dx = \int \left[\frac{\partial h}{\partial x}\right] dx = \int \left[\, -\frac{1}{K}(\, C + Dx\,)\right] dx$$
$$= -\frac{Cx}{K} - \frac{D x^2}{2K} + Const_1 \qquad (6.68)$$

(where we have substituted (6.56) for $\partial h/\partial x$), and

$$\int Q(x,y)\, dy = \int \left[\frac{\partial h}{\partial y}\right] dy = \int \left[\, \frac{1}{K}(\, B + Dx\,)\right] dy$$
$$= \frac{By}{K} + \frac{D y^2}{2K} + Const_2 \qquad (6.69)$$

(where we have substituted (6.57) for $\partial h/\partial y$). We can combine these two expression using (6.66), and at the same time, without loss of generality, combine the two constants into one, that we denote by h_o, and which we will use to denote a constant value for the hydraulic head at some reference point (x_o,y_o). Our resulting expression becomes

$$h(x,y) = h_o + \frac{By - Cx}{K} - \frac{D}{2K}\left(x^2 - y^2 \right) \qquad (6.70)$$

where B, C and D are quite arbitrary, except for being the same values appearing in (6.49). For convenience, we might make the reference point for h_o the origin $(x_o,y_o) = (0,0)$, but another point might be more appropriate for a given problem. Moreover, again without loss of generality, we might set the constant term A in (6.49) to the value of some reference flow-line V_o, so that (6.49) becomes

$$V(x,y) = V_o + Bx + Cy + Dxy \qquad (6.71)$$

The constant V_o could also be referenced to the origin, but some other point might be more appropriate, depending on the problem.

To summarize our discussion in this section, we have derived a function for $h(x,y)$ given by (6.70) consistent with our assumed form for the flow function $V(x,y)$ given by (6.49) or (6.71).

Verifying the Form of $h(x,y)$

As one plows through a theoretical derivation, it is usually best to pause occasionally and to "shoot-a-back-azimuth" to obtain some kind of conceptual check on the correctness of our computation. One would certainly hope that the algebra leading up to (6.70) is correct, and now is the time one would check it thoroughly step-by-step. Beyond that, does the form of (6.70) make sense, intuitively? Yes it does: considering that the forms for q_x and q_y are linear or first-order functions of space, one would need to differentiate a second-order function of $h(x,y)$ to obtain them. So it should be somewhat satisfying that $h(x,y)$ is a second-order function of both x and y.

Another condition that should be satisfied is that $h(x,y)$ should be a solution to the two-dimensional form of Laplace's equation

$$\frac{\partial^2 h}{\partial x^2} + \frac{\partial^2 h}{\partial y^2} = 0 \qquad (6.72)$$

which, will be recalled, implicitly involves the facts that **q** should have zero divergence (i.e. flux is conserved), and that **q** is derived from a potential $h(x,y)$; both conditions that have already been established for the present problem. So, satisfying (6.72) should not yield any new constraints on the form of $h(x,y)$; it should only verify that, in fact, the form is correct. Is it?

To check the latter, we differentiate $h(x,y)$ given by (6.70) twice with respect to x, obtaining

$$\frac{\partial^2 h}{\partial x^2} = - \frac{2D}{2K} \qquad (6.73)$$

Then differentiating $h(x,y)$ given by (6.70) twice with respect to y, we obtain

$$\frac{\partial^2 h}{\partial y^2} = + \frac{2D}{2K} \qquad (6.74)$$

Obviously, if we add (6.73) and (6.74), the sum is zero, proving that Laplace's equation (6.72) is satisfied.

Associated Flow-Net

Figure 6.8 shows the flow-net that is generated by the superposition of the two fields h(x,y) and V(x,y) from (6.70) and (6.71), respectively. The same constants appear in these two expressions, but otherwise can be assigned quite arbitrarily. For specific applications, these constants will be determined by the appropriate boundary conditions. For the case shown here, we set B = -0.05, C = 2.0, and D = -0.015. The two functions, h(x,y) and V(x,y), were then gridded for equally spaced values of x and y, and the gridded values were then contoured using a standard computer application.

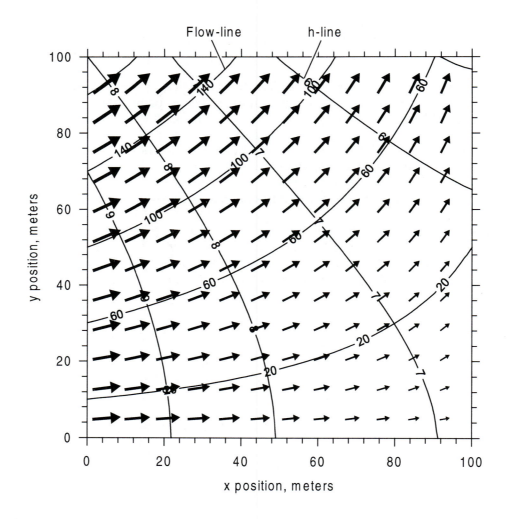

Figure 6.8 The flow-net generated by the superposition of the two fields h (x,y) and V(x,y) from (6.70) and (6.71), respectively. Superimposed is a set of arrows showing the magnitude and direction of specific discharge vectors calculated from Darcy's law.

For the case shown in Figure 6.8, x and y were calculated at 5 meter increments for 21 points along each axis, for a total of 441 points over the complete 100 by 100 meter grid. Contour intervals were selected for the h-lines and flow-lines so that the quadrilateral elements were approximately equidimensional. We have also superimposed a set of arrows showing the magnitude and direction of the local specific discharge vectors. Note that the flux follows the local trend of the flow-lines, and is strongest where the flow-lines are closest together. This is also where the orthogonal h-lines are closest together, showing that the magnitude of the local hydraulic gradient is strongest.

Constructing Flow-Nets by Hand

For many years before the advent of computers, and even today for some applications, flow-nets were, or are, constructed by hand following a few simple, but essential rules (Freeze and Cherry, 1979; Fetter, 1994; and Watson and Burnett, 1993).

1) It is essential to sketch the geometry of the problem with identical vertical and horizontal scales.

2) Indicate all explicit and implicit boundary conditions directly on the figure (such as no-flow boundaries, prescribed values of constant head, position of the water table if known).

3) By trial and error, begin sketching in flow-lines from a region of high hydraulic head to a region of low hydraulic head. *Flow-lines* should be drawn perpendicular to *h-lines* and parallel to no-flow, or impermeable, surfaces.

4) Sketch in a set of h-lines orthogonal to the flow-lines, parallel to constant head boundaries, and perpendicular to impermeable boundaries.

5) Modify the shapes of the h-lines and flow-lines until a set of approximately equidimensional quadrilateral elements is achieved.

These preliminary hand sketches are often sufficient for one's purpose in a qualitative sense; for example as a prelude to more refined modeling using another class of applications such as a numerical computer model. However, even hand constructions, if done carefully, can lead to quantitative results.

Chapter 7. Refraction of Flux

CONDITIONS ON FLOW ACROSS A DISCONTINUITY IN MATERIAL PROPERTIES

Continuity of Flux Normal to a Boundary

Consider the flux or specific discharge flowing through a small mathematical volume, shaped like a cylindrical pill box, straddling a discontinuity in material properties as shown in Figure 7.1. While the results of this section can be applied to three dimensions, we will assume only 2-D flow to simplify notation. We arrange our coordinate system such that the x direction is perpendicular or normal to the boundary, and the y direction is parallel or tangent to the boundary.

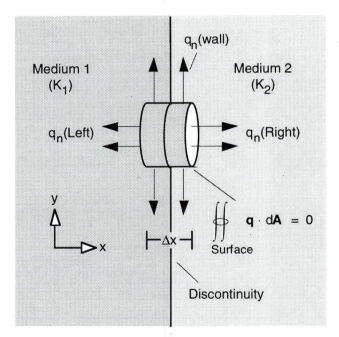

Figure 7.1 A small, cylindrical reference volume used to evaluate normal flux conditions at a discontinuity in material properties.

In the absence of sources at the interface and throughout the cylindrical volume, we assume conservation of flux according to

$$\oiint_{\text{Surface}} \mathbf{q} \cdot d\mathbf{A} = 0 \tag{7.1}$$

which we decompose into the three contributions

$$\oiint_{\text{Surface}} \mathbf{q} \cdot d\mathbf{A} = \iint_{\substack{\text{LEFT} \\ \text{FACE}}} \mathbf{q} \cdot d\mathbf{A} + \iint_{\substack{\text{RIGHT} \\ \text{FACE}}} \mathbf{q} \cdot d\mathbf{A} + \iint_{\text{WALL}} \mathbf{q} \cdot d\mathbf{A} = 0 \tag{7.2}$$

The three integrals on the right hand side of (7.2) represent the outward flux through the left face of the cylinder, through the right face of the cylinder, and through its curved cylindrical walls. We assume that the volume is mathematically small so that the normal flux is essentially uniform over each face, respectively. Thus

$$\iint\limits_{\substack{\text{LEFT}\\\text{FACE}}} \mathbf{q} \cdot d\mathbf{A} = - q_x^{(-)} \pi r_o^2 \tag{7.3}$$

$$\iint\limits_{\substack{\text{RIGHT}\\\text{FACE}}} \mathbf{q} \cdot d\mathbf{A} = q_x^{(+)} \pi r_o^2 \tag{7.4}$$

where the superscript "-" and "+" imply that the flux is evaluated, respectively, just to the left and just to the right of the discontinuity; and r_o is the radius of the cylindrical volume, so that the cross-sectional area of each face is πr_o^2. The negative sign on the right side of (7.3) is due to the outward normal of the left face being directed in the negative y direction.

The surface area of the wall of this cylindrical volume is its circumference times its height, or $2\pi r_o \Delta x$. Therefore, in the limit of Δx becoming vanishingly small, the surface area of the wall becomes vanishingly small, so that

$$\underset{\Delta x \to 0}{\text{limit}} \left[\iint\limits_{\text{WALL}} \mathbf{q} \cdot d\mathbf{A} \right] = 0 \tag{7.5}$$

Thus, in the limit when the pill box reduces to the dimension of two discs separated an infinitesimal distance to either side of the discontinuity, (7.2) reduces to

$$\oiint\limits_{\text{Surface}} \mathbf{q} \cdot d\mathbf{A} = - q_x^{(-)} \pi r_o^2 + q_x^{(+)} \pi r_o^2 = 0 \tag{7.6}$$

whence

$$q_x^{(-)} = q_x^{(+)} \tag{7.7}$$

or, in terms of an equivalent notation

$$q_x{}^{\text{Left}} = q_x{}^{\text{Right}} \tag{7.8}$$

Either (7.7) or (7.8) is a statement that the flux (or specific discharge) directed normal to a boundary is continuous.

Of course, if the normal flux is *continuous* then the normal derivative of h must be *discontinuous*, as seen from the following. According to Darcy's law, we have immediately to the left of the interface

$$q_x^{(1)} = - K_1 \frac{\partial h}{\partial x}^{\text{LEFT}} \tag{7.9}$$

whereas immediately to its right

$$q_x{}^{(2)} = - K_2 \frac{\partial h}{\partial x}^{\text{RIGHT}}$$

(7.10)

From (7.8), these latter two expressions must be equal, so that

$$K_1 \frac{\partial h}{\partial x}^{\text{LEFT}} = K_2 \frac{\partial h}{\partial x}^{\text{RIGHT}}$$

(7.11)

which can be rearranged to the form

$$\frac{\partial h / \partial x \,|_{\text{LEFT}}}{\partial h / \partial x \,|_{\text{RIGHT}}} = \frac{K_2}{K_1}$$

(7.12)

emphasizing that the derivative of h in a direction normal to an interface is discontinuous by the inverse ratio of the hydraulic conductivities. In other words, the normal derivative is smaller in media where the conductivity is larger, and vice versa.

Continuity of h

Another constraint on the subsurface flow regime is that the hydraulic head is a continuous function of position, even across finite discontinuities in material properties. This condition is a consequence of h being a surrogate for the total energy W of a packet of fluid. The relation between h and W was discussed in some detail in an earlier chapter, and the conservation of energy imposes significant constraints on the mathematical properties of the hydraulic head, not the least of which is that h, itself, can be viewed as a potential function — and potential functions are continuous functions of space, even across finite discontinuities in the material properties of a medium. This fact (along with the continuity of normal flux discussed in the last section) places strong constraints on the physical and mathematical inter-relationships between the flow fields and potential (h) fields on either side of a discontinuity. This section will discuss in some detail the basis for this assertion and its implications.

Note: We are *not* implying that all the *derivatives* of h need to be continuous across an interface. The last section showed, in fact, that the normal derivative of h is expected to be discontinuous across a discontinuity in material properties. We *are* saying, however, that $h(x,y)$, itself, is continuous.

Continuity of h: Argument 1. The first argument for the continuity of $h(x,y)$ is basically intuitive, and is couched in physical terms. If one attempted to conceive of a situation where the function h *did* indeed undergo a finite discontinuity Δh over a vanishing distance $\delta l \rightarrow 0$ (but where the hydraulic conductivity K remained finite), then Darcy's law would imply a non-physical singularity in the flux, since $q = -K \Delta h / \delta l \rightarrow \infty$. Since such singular flows do not occur (unless exceptional constructions are physically or mathematically fabricated at the interface), we assume that $\Delta h = 0$, thus that $h(x,y)$ is continuous.

Continuity of h: Argument 2. The second argument for the continuity of $h(x,y)$ might be considered to be more mathematical, but in fact builds on the theme that h is a potential function.

Recall that one aspect of potential functions is that line integrals of the form

$$\int_A^B dh = \int_A^B \left[\frac{\partial h}{\partial x} dx + \frac{\partial h}{\partial y} dy \right] \tag{7.13}$$

are independent of path, and depend only on the value of h at the end-points A and B, such that

$$\int_A^B dh = h_B - h_A \tag{7.14}$$

Consider in detail a path A-B perpendicular to the interface, as shown in Figure 7.2.

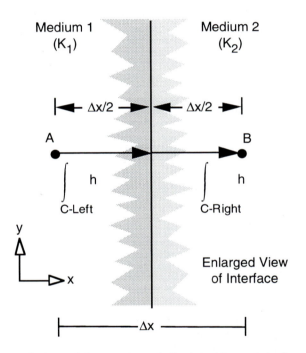

Figure 7.2 An enlarged view of the contour A-B crossing an interface between two media having hydraulic conductivities of K_1 and K_2, respectively.

We decompose this branch into a left and a right element according to

$$\int_A^B dh = \int_{C\text{-Left}} dh + \int_{C\text{-Right}} dh$$

$$= \int_{C\text{-Left}} \left[\frac{\partial h}{\partial x} dx + \frac{\partial h}{\partial y} dy \right] + \int_{C\text{-Right}} \left[\frac{\partial h}{\partial x} dx + \frac{\partial h}{\partial y} dy \right] \tag{7.15}$$

Since these paths lie in the x direction, the differential dy = 0, so that (7.15) reduces to

$$\int_A^B dh = \int_{C\text{-Left}} \frac{\partial h}{\partial x} dx + \int_{C\text{-Right}} \frac{\partial h}{\partial x} dx \qquad (7.16)$$

For mathematically small distances where a function can be represented by a first order Taylor's series expansion at a point, we may use Darcy's law to write

$$\int_{C\text{-Left}} \frac{\partial h}{\partial x} dx = \frac{\Delta x}{2} \left[\frac{\partial h}{\partial x} \bigg|_{x = -\Delta x/2} \right] = -\frac{\Delta x}{2} \frac{q_x^{\text{Left}}}{K_1} \qquad (7.17)$$

and

$$\int_{C\text{-Right}} \frac{\partial h}{\partial x} dx = \frac{\Delta x}{2} \left[\frac{\partial h}{\partial x} \bigg|_{x = +\Delta x/2} \right] = -\frac{\Delta x}{2} \frac{q_x^{\text{Right}}}{K_2} \qquad (7.18)$$

From our earlier discussion on the continuity of normal flux at the interface, we have

$$q_x^{\text{Left}} = q_x^{\text{Right}} \qquad (7.8)$$

so that, in the limit of small Δx, the integral becomes

$$\int_A^B dh = -\frac{\Delta x}{2} \left(\frac{1}{K_1} + \frac{1}{K_2} \right) q_x^{\text{Interface}} \qquad (7.19)$$

Upon substituting the right hand side of (7.14)

$$\int_A^B dh = h_B - h_A \qquad (7.14)$$

for the left hand side of (7.19), we obtain

$$h_B - h_A = -\frac{\Delta x}{2} \left(\frac{1}{K_1} + \frac{1}{K_2} \right) q_x^{\text{Interface}} \qquad (7.20)$$

Assuming K_1, K_2 and q_x at the interface are finite, in the limit of vanishing Δx the right hand side of (7.20) becomes vanishingly small. Hence, at the interface

$$h_A = h_B \qquad (7.21)$$

demonstrating that h is continuous at the interface.

A consequence of (7.21) is the following condition on the partial derivative of h with respect to y at the interface

$$\frac{\partial h}{\partial y}^{RIGHT} = \frac{\partial h}{\partial y}^{LEFT} \tag{7.22}$$

This is a statement that the derivative of h in a direction *tangential* to an interface is continuous across the interface.

If the tangential derivative is continuous, then from Darcy's law,

$$\frac{\partial h}{\partial y}^{LEFT} = - \frac{q_y^{(1)}}{K_1} \tag{7.23}$$

and

$$\frac{\partial h}{\partial y}^{RIGHT} = - \frac{q_y^{(2)}}{K_2} \tag{7.24}$$

From (7.22), we may equate the right sides of (7.23) and (7.24), to obtain

$$\frac{q_y^{(1)}}{K_1} = \frac{q_y^{(2)}}{K_2} \tag{7.25}$$

which can be rearranged to the form

$$\frac{q_y^{(1)}}{q_y^{(2)}} = \frac{K_1}{K_2} \tag{7.26}$$

showing that the *tangential* component of flux is *discontinuous* across an interface by the ratio of the hydraulic conductivities.

Summary of Conditions on q and h at a Discontinuity

We conclude this section by summarizing the following principal conditions on flow at an interface:

    1) The flux normal to an interface is continuous
    2) h is continuous across an interface

with the corollaries:

    3) The tangential derivative of h is continuous [a corollary to (2)]
    4) The normal derivative of h is discontinuous by the ratio of hydraulic conductivities [a corollary to (1)]
    5) The flux tangential to an interface is discontinuous by the ratio of hydraulic conductivities [a corollary to (2)]

REFRACTION OF FLOW ACROSS A DISCONTINUITY

Consider the behavior of flow near a discontinuity in the hydraulic conductivity (Figure 7.3).

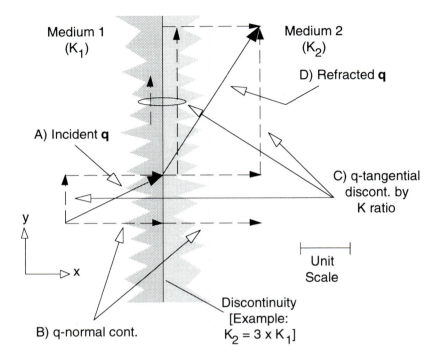

Figure 7.3 Details on the refraction of a flux line in the vicinity of a discontinuity. For the case shown, the conductivity contrast is $K_2 = 3 \times K_1$.

This example is carefully scaled for a conductivity contrast of $K_2 = 3 \times K_1$ such that, with reference to the labels in the figure:

A) The incident flux $\mathbf{q}_1$ has components $(q_x, q_y) = (2, 1)$;

B) The normal flux, q_x, is continuous across the interface (.i.e. has the same magnitude, 2 units, and direction on either side of the boundary);

C) q_y is discontinuous by the ratio of the conductivities, $K_2 / K_1 = 3$;

D) The consequent refraction of $\mathbf{q}_2$ as shown, away from the normal.

Figure 7.4 shows the behavior of the field lines of hydraulic head in the vicinity of the discontinuity. The notation is such that the subscript refers to the contour number of the respective h-line; for the direction of flow indicated we must have $h_{n+1} < h_n$. For example $h_1 = 110$ meters, $h_2 = 108$ meters, $h_3 = 106$ meters, and so forth. The superscript (1) or (2) corresponds to the respective medium on either side of the interface, but of course, because of continuity of h

$$h_1^{(1)} = h_1^{(2)}$$
$$h_2^{(1)} = h_2^{(2)}$$
$$(\dots)$$

The "kink" in each h-line is due to a discontinuity in the normal derivative of h (because q-normal is continuous).

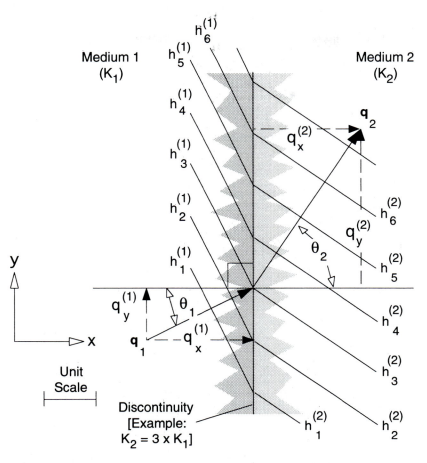

Figure 7.4 The same example as Figure 7.3, but showing the continuity of the field lines of hydraulic head across the discontinuity.

Decomposing the total flux on each side of the discontinuity into its respective components, and simplifying the figure somewhat, we obtain Figure 7.5. From simple trigonometry

$$q_y^{(1)} = q_x^{(1)} \tan(\theta_1) \tag{7.27}$$

and

$$q_y^{(2)} = q_x^{(2)} \tan(\theta_2) \tag{7.28}$$

Dividing (7.27) by (7.28), we obtain

$$\frac{q_y^{(1)}}{q_y^{(2)}} = \frac{q_x^{(1)} \tan(\theta_1)}{q_x^{(2)} \tan(\theta_2)} \tag{7.29}$$

From continuity of q perpendicular to the boundary, we have

$$q_x^{(1)} = q_x^{(2)}$$ (7.30)

which, substituted into the right side of (7.29), reduces the expression to

$$\frac{q_y^{(1)}}{q_y^{(2)}} = \frac{\tan(\theta_1)}{\tan(\theta_2)}$$ (7.31)

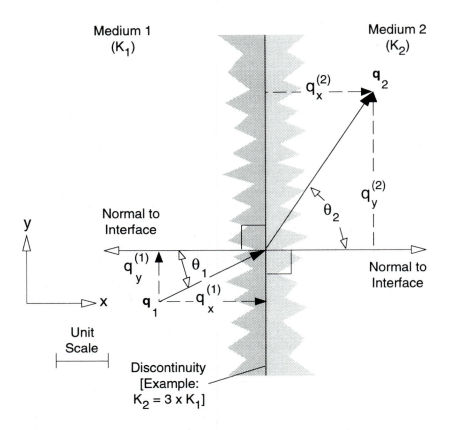

Figure 7.5 Relation between angle of incidence and angle of refraction across a hydraulic interface.

Next, we recall that the continuity of h along and across the boundary results in

$$h^{(1)} = h^{(2)}$$ (7.32)

which is, in turn, equivalent to

$$\frac{\partial h^{(1)}}{\partial y} = \frac{\partial h^{(2)}}{\partial y}$$ (7.33)

across the boundary.

Substituting from Darcy's law into the last relation, we have

$$\frac{q_y^{(1)}}{K_1} = \frac{q_y^{(2)}}{K_2} \qquad (7.34)$$

which can be rearranged to the form

$$\frac{q_y^{(1)}}{q_y^{(2)}} = \frac{K_1}{K_2} \qquad (7.35)$$

which, as discussed above, is a statement on the discontinuity in the tangential component of flux at an interface.

Substituting the right hand side of (7.35) for the left hand side of (7.31) above, we obtain the *tangent law of hydraulic refraction*:

$$\frac{K_1}{K_2} = \frac{\tan(\theta_1)}{\tan(\theta_2)} \qquad (7.36)$$

A Numerical Example

It may be instructive to compare the tangent law against the geometrical constructions of Figures 7.4 and 7.5. For the example in these figures, the incident flux line $\mathbf{q}_1$ was draw so that

$$q_x^{(1)} = 2 \text{ units}$$

and

$$q_y^{(1)} = 1 \text{ unit}$$

which leads to an incident angle of flux of

$$\theta_1 = \tan^{-1}\left(\frac{1}{2}\right) = 26.6^{\circ}$$

For the example shown in the figures, we have set

$$\frac{K_1}{K_2} = \frac{1}{3}$$

so that from the law of tangents

$$\theta_2 = \tan^{-1}\left(\frac{K_2}{K_1}\tan(\theta_1)\right) = \tan^{-1}\left(\frac{3}{1}\tan(26.5^{\circ})\right)$$

or

$$\theta_2 = 56.3^{\circ}$$

which compares well to the geometrical construction in Figures 7.4 and 7.5 (within a drafting uncertainty of $\pm 0.5^o$). This angle could have been calculated directly from the continuity conditions on h and the normal flux at the interface. For the case shown here, where

$$q_x^{(1)} = 2 \text{ units} \quad \text{and} \quad q_y^{(1)} = 1 \text{ unit}$$

from the continuity of normal **q**, we would expect:

$$\text{If} \quad q_x^{(1)} = 2 \text{ units}, \quad \text{then} \quad q_x^{(2)} = 2 \text{ units}$$

In addition, from the continuity of h across the interface, we would also expect the continuity of the tangential derivative of h, or $\partial h / \partial y$, across the interface, which from Darcy's law and the conductivity contrast assumed here, $K_2 = 3 \times K_1$, leads to the following condition on

$$\text{If} \quad q_y^{(1)} = 1 \text{ unit}, \quad \text{then} \quad q_y^{(2)} = 3 \text{ units}$$

Thus from simple trigonometry

$$\theta_2 = \tan^{-1}\left(\frac{q_y^{(2)}}{q_x^{(2)}}\right) = \tan^{-1}\left(\frac{3}{2}\right) = 56.3^o$$

as before. This simply reaffirms that the law of tangents is a concise statement of continuity conditions at a discontinuity in material properties. It is much more important to understand the physical basis for such relationships than it is to memorize such formulae for their own sake.

Chapter 8. Steady State Flow in Unconfined Aquifers

FUNDAMENTAL ASPECTS OF UNCONFINED FLOW

The local hydraulic head of an unconfined aquifer is the local elevation of the water table — a free surface that can adjust in elevation to various flow conditions. Often unconfined aquifers are referred to as "water table aquifers".

In the absence of local sources, flow does not cross this free surface so that the water table physically represents a flow contour, or a flow surface. In such a case, the local hydraulic gradient is tangential to the water table, and the negative gradient points in the direction of local flow as illustrated schematically in Figure 8.1.

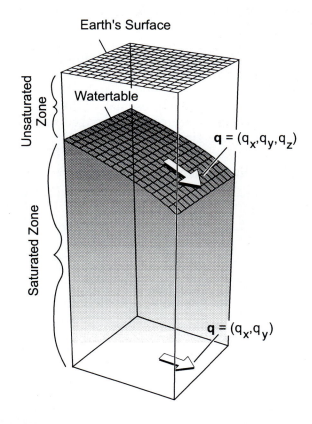

Figure 8.1 Three dimensional schematic illustrating unconfined flow in a uniform medium.

Thus, at and below the water table, fluid flow is usually three dimensional

$$\mathbf{q}(x,y,z) = q_x(x,y,z)\widehat{\mathbf{x}} + q_y(x,y,z)\widehat{\mathbf{y}} + q_z(x,y,z)\widehat{\mathbf{z}} \tag{8.1}$$

where, as before, $\widehat{\mathbf{x}}$, $\widehat{\mathbf{y}}$ and $\widehat{\mathbf{z}}$ are unit vectors.

At, and beneath, the water table the hydraulic head is generally of the form h(x,y,z), and for those specific values of x',y' and z' where

$$h(x',y',z') = \text{constant} = h_c \qquad (8.2)$$

one generates a surface (our three dimensional version of an h-line in 2-D) over which h is everywhere equal to the same value h_c.

Commensurate with the water table being a local flow surface, local h-lines in 2-D, or h-surfaces in 3-D, are orthogonal to the local water table, *providing there are no local sources or sinks of fluid to cause flow across the water table surface*. If, in fact, there *are* local sources or sinks of fluid, the local water table is no longer a flow-line or flow surface. Typical *sources* might be infiltration from natural precipitation or man-made leach-fields; typical *sinks* might be evaporation or transpiration through plant and tree roots, seepage surfaces and so forth. In either case, sources or sinks cause a flux across the water table, so that the concept of the water table being a flow-line that bounds fluid flow is violated.

For now, however, we assume source-free and sink-free conditions, and assume that the water table is a local flow-line in 2D or flow contour in 3D.

We assume that the total flow in an unconfined aquifer of uniform conductivity K is bounded by the water table at the top and an impermeable layer at the bottom. To simplify our discussion, we will also assume that this lower boundary is horizontal, and is the common datum level to which we refer measurements of hydraulic head. A representative flow net for unconfined flow is shown in vertical section in Figure 8.2, where the h-lines are arranged such that $h_n > h_{n+1}$. Also shown are flow-lines, with arrows showing the direction of flow.

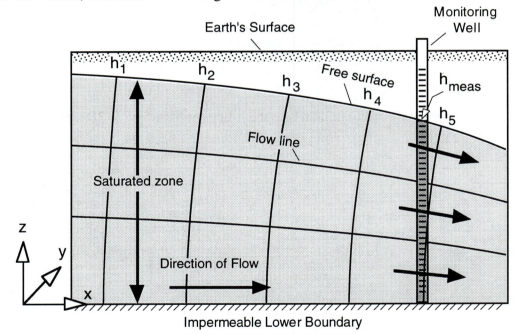

Figure 8.2 A vertical cross-section in the plane of flow illustrating unconfined flow in a uniform medium. The free surface, or water table, is a flow-line for source-free conditions, and h-lines intercept it orthogonally.

In most cases of unconfined flow, the hydraulic gradient has three spatial components

$$\text{grad } h = \nabla h = \frac{\partial h}{\partial x} \hat{\mathbf{x}} + \frac{\partial h}{\partial y} \hat{\mathbf{y}} + \frac{\partial h}{\partial z} \hat{\mathbf{z}} \tag{8.3}$$

For strictly horizontal flow, as in horizontal confined aquifers, $\partial h/\partial z = 0$, so that h is invariant with z. In the case considered in Figure 8.2, *there will be a small vertical component of flow*, so that h will be dependent on z. This is illustrated in Figure 8.3 where the coordinate system has been oriented so that x is pointing in the direction of flow — the water table will accordingly have a slight dip toward increasing x, and the field line that starts at z = 0 beneath the surface point (x_0, y_0), with increasing z tends to have an increasing deflection in the direction of flow.

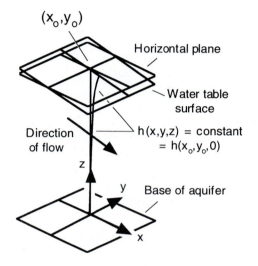

Figure 8.3 Local deflection from vertical of an h-line in unconfined flow.

The deflection is greatest right at the water table where the h-line (or h-surface in 3D) intercepts the free surface at right angles (under source-free conditions).

Consider a monitoring well screened (or slotted) throughout the entire section of the aquifer shown in Figure 8.2. Because the well casing is open throughout its length, the elevation of water in the well is at the level of the local free surface, which is to say the water table. We might denote this on a plan map of the area as the local head $h(x_5, y_5)$, which we might define as the elevation of the water table beneath a horizontal position of (x_5, y_5). However, this is generally not the value of the hydraulic head at the footprint of the h field line originating at z = 0 beneath this point, which we have denoted in Figure 8.2 by h_5.

This may be clearer in Figure 8.4, which shows an enlarged view of the portion of Figure 8.2 in the vicinity of the monitoring well.

Typically for unconfined flow, at an arbitrary point on the earth's surface (x_0, y_0),

$$h(x_0, y_0, h) \neq h(x_0, y_0, z_0) \tag{8.4}$$

As shown in Figure 8.4, we denote this difference by δh.

Clearly, there is a systematic difference between the head at a point on the water table directly over a point (x,y) and the head at the base of the aquifer at z = 0.

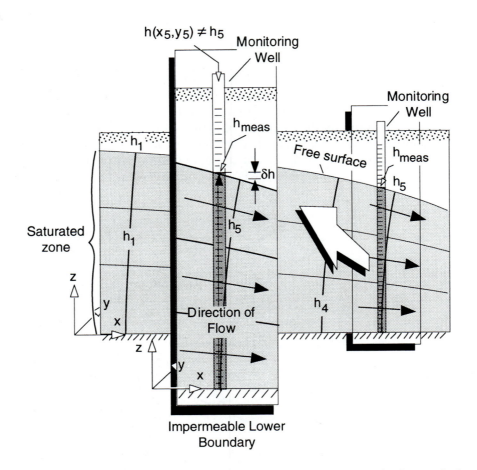

Figure 8.4 An enlarged view of conditions in the vicinity of the monitoring well shown in Figure 3.2. Note the small but finite difference δh between h measured in the well and h_5, the h-line whose footprint originates at z = 0 beneath the well.

DUPUIT FLOW IN UNCONFINED AQUIFERS

Dupuit Assumptions

While the last section emphasized that unconfined flow is generally three dimensional even in the simplest cases, the dependence of h on z is often a *weak* dependence. Clearly, for the case illustrated in Figures 8.2 and 8.4, the difference is small between the head at a point on the water table directly over (x,y) and the head at the point where $h(x,y,z) = h_c$, a constant, actually intercepts the free surface.

What do we mean by "small"? A reasonable measure would be that the difference, δh, is small relative to z = h_{avg} in the vicinity of (x,y) as shown in Figure 8.5, where the circumflex over each h indicates a a value for the hydraulic head determined by a spatial average over an appropriate local averaging volume.

If so, we can assume that h at the surface of the aquifer directly above (x,y) [at z = h] is a close approximation to the hydraulic head at the base of the aquifer [at z = 0]. In other words

$$h(x,y,h) \cong h(x,y,0). \tag{8.5}$$

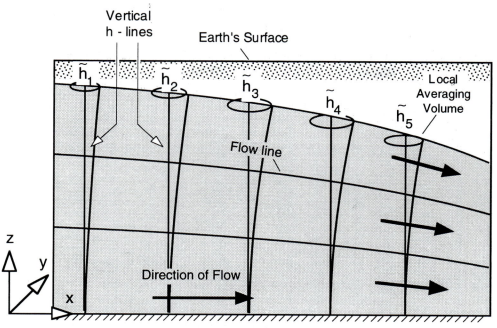

Figure 8.5 Showing the presumed relationship between the spatially averaged hydraulic head over a local radius about each well (shown by the circles), and the approximation of the local head over the thickness of the aquifer by a vertical line, or, in 3-D, a vertical plane.

This is equivalent to assuming $\partial h/\partial z \cong 0$, and that h is invariant with z throughout the aquifer — or stated in another way, lines or surfaces of constant h are approximately vertical. The more powerful (but less valid) statement

$$h(x,y,h) \equiv h(x,y,0). \tag{8.6}$$

is known among hydrologists as the basis for the *Dupuit assumptions* regarding unconfined flow. These are named for the French scientist who, in 1863, laid the groundwork for a theory that, even today, significantly improves the tractability of mathematical solutions to certain classes of unconfined flow, several of which are discussed below.

Average Flow Properties of the Aquifer

If one is primarily interested in the large scale flow properties of an unconfined aquifer (on a regional scale, for example, or even over its total thickness) it is convenient to discuss these properties in an *average* sense. For example, the "transmissivity" of a unconfined aquifer is the depth integrated hydraulic conductivity, and is given by

$$T = \int_0^h K\,dz \tag{8.7}$$

If we divide the transmissivity by the local thickness of the saturated section — the hydraulic head h, the local elevation of the water table — we obtain the *average hydraulic conductivity* for the formation

$$\widetilde{K} = \langle K \rangle = \frac{1}{h} \int_0^h K \, dz \tag{8.8}$$

The total horizontal discharge in a particular direction (say the Q_x component) is given by the depth integrated specific discharge

$$Q_x = \int_0^h \mathbf{q} \cdot \hat{\mathbf{x}} \, dz \tag{8.9}$$

Again, dividing by the local thickness of the saturated section — the hydraulic head h — we obtain the *average specific discharge*, for the average horizontal x component

$$\widetilde{q_x} = \langle q_x \rangle = \frac{1}{h} \int_0^h \mathbf{q} \cdot \hat{\mathbf{x}} \, dz \tag{8.10a}$$

and for the average horizontal y component would have the form

$$\widetilde{q_y} = \langle q_y \rangle = \frac{1}{h} \int_0^h \mathbf{q} \cdot \hat{\mathbf{y}} \, dz \tag{8.10b}$$

In addition, we make the simplifying Dupuit assumption that all flow is horizontal, and therefore that $\widetilde{q_z} = \langle q_z \rangle \equiv 0$.

Theoretical and Practical Basis for Dupuit Flow

For the remainder of this text, unless explicitly stated otherwise, when discussing unconfined flow we will assume that K and q_x and q_y represent these average parameters, that lines or surfaces of constant hydraulic head h are vertical, and that on the average q_z and $\partial h/\partial z$ are negligible. These are the underpinning assumptions and implied consequences of the Dupuit theory (also known as the Dupuit-Forscheimer theory by some authors) of unconfined or free-surface flow. We will refer to flow that can be approximated by these conditions as "Dupuit flow" (c.f. Freeze and Cherry, 1979; Fetter, 1994).

From a practical view, de Marsily (1986) argues that the Dupuit assumptions are quite consistent with field observations, providing one is not too close to groundwater divides or to outlet zones where there can be substantial flow in the vertical direction. From a theoretical viewpoint, Freeze and Cherry (1979) point out that more rigorous calculations verify the adequacy of the Dupuit assumptions providing that hydraulic gradients, more specifically the slope of the free surface, is small. As we will see below, the computational advantage of the Dupuit theory is substantial when the appropriate assumptions can be applied, but in Chapter 9 we will discuss a counter-example in which flow in non-Dupuit.

HORIZONTAL UNCONFINED FLOW: DIVERGENCE OF FLUX FOR THE 1-D CASE

Consider the case illustrated below of unconfined or water table flow only in the vertical x-z plane. We apply the principle of conservation of mass flux

$$\oiint_{surface} \mathbf{q} \cdot d\mathbf{A} = 0 \qquad (8.11)$$

to one-dimensional Dupuit flow through the rectilinear prism as illustrated below. (Remember that the h-lines are assumed to be vertical for Dupuit flow.)

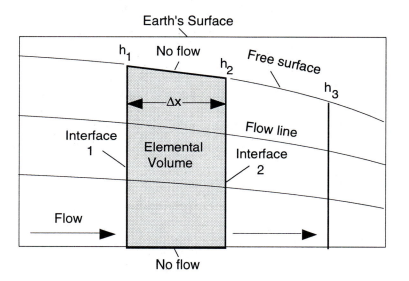

Figure 8.6 Elemental volume used to describe the conservation of Dupuit flow.

This results in

$$\int_0^{h_2} q_x^{(2)} \, dz - \int_0^{h_1} q_x^{(1)} \, dz = 0 \qquad (8.12)$$

where, as shown in Figure 8.7, $q_x^{(1)}$ and $q_x^{(2)}$ represent the average specific discharges across interface 1 and 2, respectively.

Note: Although there may be a vertical component of flow beneath the surface of the water table, no flux passes through the top surface (the water table which in this case is a flow-line), nor the lower surface (the impermeable lower boundary). It is only the *horizontal* component q_x that contributes to the total inflow and outflow of flux through the vertical walls of this prism.

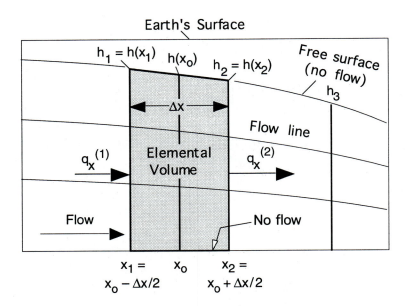

Figure 8.7 Showing normal flux into and out of an elemental volume.

Since $q_x^{(1)}$ and $q_x^{(2)}$ represent the *average* specific discharge over the vertical dimension of the aquifer, each can be brought outside its respective integral, and the integrals simplified to the form

$$h_2 q_x^{(2)} - h_1 q_x^{(1)} = 0 \qquad (8.13)$$

From Darcy's law, we have

$$q_x^{(1)} = - K \left. \frac{\partial h}{\partial x} \right|_{x = x_1} \qquad (8.14)$$

and

$$q_x^{(2)} = - K \left. \frac{\partial h}{\partial x} \right|_{x = x_2} \qquad (8.15)$$

Substituting (8.15) and (8.14) into (8.13) leads to

$$h_2 \, K \, \frac{\partial h}{\partial x}^{(2)} - h_1 \, K \, \frac{\partial h}{\partial x}^{(1)} = 0 \qquad (8.16)$$

Recall the Taylor's series expansion for $h(x)$

$$h(x) = h(x_o) + (x - x_o) \left. \frac{\partial h}{\partial x} \right|_{x_o} + \frac{(x-x_o)^2}{2} \left. \frac{\partial^2 h}{\partial x^2} \right|_{x_o} + \dots \qquad (8.17)$$

To first order, at the left and right vertical facet, respectively, we have

$$h_1 = h(x_o - \frac{\Delta x}{2}) = h(x_o) - \frac{\Delta x}{2} \left. \frac{\partial h}{\partial x} \right|_{x_o} \qquad (8.18)$$

and

$$h_2 = h(x_o + \frac{\Delta x}{2}) = h(x_o) + \frac{\Delta x}{2} \frac{\partial h}{\partial x}\bigg|_{x_o} \qquad (8.19)$$

where $h(x_o)$ is the elevation of the free surface at the mid-point of our elemental volume, and Δx is the width of our elemental volume from one side to the other.

Upon differentiating (8.17) with respect to x, and evaluating the result at the $x - x_o = -\Delta x/2$ and $x - x_o = +\Delta x/2$ facets, respectively, we obtain to first order

$$\frac{\partial h^{(1)}}{\partial x} = \frac{\partial h}{\partial x}\bigg|_{x = x_1} \approx \frac{\partial h}{\partial x}\bigg|_{x_o} - \frac{\Delta x}{2} \frac{\partial^2 h}{\partial x^2}\bigg|_{x_o} \qquad (8.20)$$

and

$$\frac{\partial h^{(2)}}{\partial x} = \frac{\partial h}{\partial x}\bigg|_{x = x_2} \approx \frac{\partial h}{\partial x}\bigg|_{x_o} + \frac{\Delta x}{2} \frac{\partial^2 h}{\partial x^2}\bigg|_{x_o} \qquad (8.21)$$

Substituting the *left* hand sides of (8.20) and (8.21) into (8.16) [we will invoke the information on the *right* hand sides shortly], we obtain

$$K\left[h_o + \frac{\Delta x}{2} \frac{\partial h}{\partial x}\bigg|_{x_o}\right] \frac{\partial h^{(2)}}{\partial x} - K\left[h_o - \frac{\Delta x}{2} \frac{\partial h}{\partial x}\bigg|_{x_o}\right] \frac{\partial h^{(1)}}{\partial x} = 0 \qquad (8.22)$$

Dividing through by the common factor Δx, we rearrange the expression to the form

$$K h_o \left[\frac{\left(\frac{\partial h^{(2)}}{\partial x} - \frac{\partial h^{(1)}}{\partial x}\right)}{\Delta x}\right] + \frac{K}{2} \frac{\partial h}{\partial x}\bigg|_{x_o} \left[\frac{\partial h^{(2)}}{\partial x} + \frac{\partial h^{(1)}}{\partial x}\right] = 0 \qquad (8.23)$$

We next invoke the approximations from the right hand sides of (8.20) and (8.21). Upon *subtracting* the right hand term of (8.20) from that of (8.21), dividing by Δx, in the limit of Δx approaching zero we obtain

$$\lim_{\Delta x \to 0} \left[\frac{\left(\frac{\partial h^{(2)}}{\partial x} - \frac{\partial h^{(1)}}{\partial x}\right)}{\Delta x}\right] = \frac{\partial^2 h}{\partial x^2}\bigg|_{x_o} \qquad (8.24)$$

and upon *adding* the right hand term of (8.20) to that of (8.21), in the limit of Δx approaching zero we obtain

$$\lim_{\Delta x \to 0} \left[\frac{\partial h^{(2)}}{\partial x} + \frac{\partial h^{(1)}}{\partial x}\right] = 2 \frac{\partial h}{\partial x}\bigg|_{x_o} \qquad (8.25)$$

Substituting the right hand sides of (8.24) and (8.25), respectively, for the appropriate terms in (8.23), we obtain

$$K\,h_o\,\frac{\partial^2 h}{\partial x^2}\bigg|_{x_o} + K\left(\frac{\partial h}{\partial x}\bigg|_{x_o}\right)^2 = 0 \qquad (8.26)$$

which, since the reference point x_o is quite arbitrary, can be written for a general x position as

$$K\,h\,\frac{\partial^2 h}{\partial x^2} + K\left(\frac{\partial h}{\partial x}\right)^2 = 0 \qquad (8.27)$$

or simply

$$K\,\frac{\partial}{\partial x}\left[h\,\frac{\partial h}{\partial x}\right] = 0 \qquad (8.28)$$

Expression (8.28) is a nonlinear second order differential equation in h for unconfined flow in one dimension.

Horizontal Unconfined Flow: The 2-D Case

For the more general case of two dimensional unconfined flow, (8.28) can be generalized to the case

$$K\,\frac{\partial}{\partial x}\left[h\,\frac{\partial h}{\partial x}\right] + K\,\frac{\partial}{\partial y}\left[h\,\frac{\partial h}{\partial y}\right] = 0 \qquad (8.29)$$

It may be useful in understanding the implications of (8.29) to think of it as a divergence condition on the depth-integrated discharge vectors Q_x and Q_y. This is done by recalling that for horizontal flow in a uniform, isotropic medium the total discharge vector is the product of the *specific* discharge vector and the *thickness* of the saturated section of the aquifer — which in the case of an unconfined aquifer has a local value of h. Thus, the total discharge components are

$$Q_x = h\,q_x \qquad (8.30)$$

and

$$Q_y = h\,q_y \qquad (8.31)$$

Substituting for q_x and q_y from Darcy's law, we obtain

$$Q_x = -\,K\,h\,\frac{\partial h}{\partial x} \qquad (8.32)$$

and

$$Q_y = -\,K\,h\,\frac{\partial h}{\partial y} \qquad (8.33)$$

Upon substituting the left hand terms in (8.32) and (8.33) for their corresponding right hand terms in (8.29), we obtain

$$\frac{\partial}{\partial x}\left[Q_x\right] + \frac{\partial}{\partial y}\left[Q_y\right] = 0 \qquad (8.34)$$

The latter can be written simply as

$$\nabla \cdot \mathbf{Q} = 0 \tag{8.35}$$

where

$$\mathbf{Q} = [Q_x(x,y), Q_y(x,y)] \tag{8.36}$$

Expressions (8.34) and (8.35) underscore the fact that (8.29) is equivalent to a simple divergence condition on the total discharge vector (8.36).

Discharge Potential for Unconfined Flow

From elementary calculus, we recall that the partial derivatives of the square of a continuous function, f, of two variables, u and v, are

$$\frac{\partial}{\partial u}\left(f(u,v)^2 \right) = 2\,f\frac{\partial f}{\partial u} \tag{8.37}$$

and

$$\frac{\partial}{\partial v}\left(f(u,v)^2 \right) = 2\,f\frac{\partial f}{\partial v} \tag{8.38}$$

If we let u be used to denote the variable x, and v to denote the variable y, then upon substituting h(x,y) for f(u,v), we obtain after rearranging

$$\frac{1}{2}\frac{\partial}{\partial x}\left(h(x,y)^2 \right) = h\frac{\partial h}{\partial x} \tag{8.39}$$

and

$$\frac{1}{2}\frac{\partial}{\partial y}\left(h(x,y)^2 \right) = h\frac{\partial h}{\partial y} \tag{8.40}$$

Thus, upon defining a new function

$$\Phi = \frac{K\,h^2}{2} + C_u \tag{8.41}$$

which we will call the *discharge potential for unconfined flow*, we see that the horizontal discharge components (8.32) and (8.33) become

$$Q_x = -\frac{\partial \Phi}{\partial x} \tag{8.42}$$

and

$$Q_y = -\frac{\partial \Phi}{\partial y} \tag{8.43}$$

so that upon substituting (8.42) and (8.43) into the divergence condition (8.29) [or (8.34)], we obtain

$$\frac{\partial^2 \Phi}{\partial x^2} + \frac{\partial^2 \Phi}{\partial y^2} = 0 \tag{8.44}$$

which is Laplace's equation in Φ. We have, in a sense, "linearized" the nonlinear form (8.29) to the linear form (8.44).

APPLICATION TO 1-D STEADY-STATE FLOW IN AN UNCONFINED AQUIFER

Flow in the Absence of Local Sources (Source-Free Conditions)

Statement of the problem. Consider one of the simplest geometries for flow in an unconfined aquifer. As illustrated below (following Fetter, 1994), assume we have a vertical 2-D section cutting through two rivers or water-filled channels separated a distance L. The height of water in Channel #1 is h_1, the height in Channel #2 is h_2. The hydraulic head of each channel is constant in time. An unconfined or water table aquifer of uniform hydraulic conductivity K connects the two. The aquifer is bounded below by a subhorizontal impermeable confining layer.

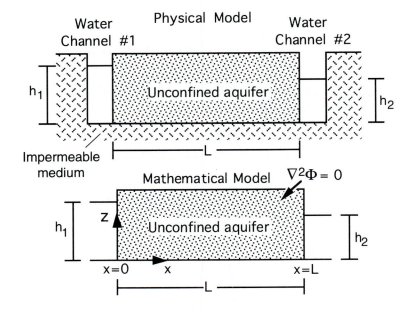

Figure 8.8 Example of a 1-D unconfined aquifer.

Solution in terms of the discharge potential. While the general relation for horizontal flow in an unconfined aquifer is given by (8.44) above, for the specific case considered here, $\partial^2\Phi/\partial y^2 = 0$, because of the 1-D nature of flow. Thus, in the interior of the aquifer, we have to solve the 1-D Laplace equation

$$\partial^2\Phi/\partial x^2 = 0 \qquad (8.45)$$

subject to the boundary conditions on the hydraulic head that $h = h_1$ at $x = 0$, and $h = h_2$ at $x = L$. According to our definition of the discharge potential for unconfined flow given by (8.31), these latter two boundary conditions impose the following corresponding conditions on Φ.

At $x = 0$

$$\Phi = \Phi_1 = (Kh_1{}^2)/2 \qquad (8.46)$$

At x = L

$$\Phi = \Phi_2 = (Kh_2^2)/2 \tag{8.47}$$

where we have set the constant C_u in (8.31) equal to 0 without loss of generality. Writing (8.45) in the form

$$\partial \, [\partial\Phi/\partial x] \, /\partial x = 0 \tag{8.48}$$

we can integrate once to obtain

$$\partial\Phi/\partial x = \text{a constant} \tag{8.49}$$

which we will set equal to A. We can now integrate

$$\partial\Phi/\partial x = A \tag{8.50}$$

to give

$$\Phi(x) = Ax + B \tag{8.51}$$

where B is a second constant of integration. The values of A and B are determined from the boundary conditions on Φ at x = 0 and x = L. At x = 0

$$\Phi(0) = A \cdot 0 + B \tag{8.52}$$

which we set equal to the boundary value at x = 0, $\Phi(0) = \Phi_1$ or

$$\Phi(0) = A \cdot 0 + B = \Phi_1 \tag{8.53}$$

so that

$$B = \Phi_1 \tag{8.54}$$

At x = L,

$$\Phi(L) = \Phi_2 = A \, L + \Phi_1 \tag{8.55}$$

which can be solved for

$$A = [\Phi_2 - \Phi_1]/L \tag{8.56}$$

Thus, our general solution (8.51) can be rewritten in terms of the above expressions for A and B, given respectively by (8.56) and (8.54), to determine the following equation for the discharge potential:

$$\Phi(x) = [(\Phi_2 - \Phi_1)/L] \, x + \Phi_1 \tag{8.57}$$

Converting to hydraulic head. It is more usual, of course, to express results in terms of the hydraulic head. Upon substituting the definition for the discharge potential for unconfined flow

$$\Phi = \frac{K \, h^2}{2} + C_u \tag{8.58}$$

into (8.57), and noting that for the specific case considered here we can set $C_u = 0$, we obtain

$$h^2(x) = [(h_2^2 - h_1^2)/L]\,x + h_1^2 \qquad (8.59)$$

or, taking the square root of both sides,

$$h(x) = \sqrt{h_1^2 + \frac{(h_2^2 - h_1^2)}{L}\,x} \qquad (8.60)$$

The functional form of the hydraulic head given by (8.60) is a parabola (lying on its side). Using $h(x)$ from (8.60), Figure 8.9 is a normalized plot of $[h(x) - h_2]/[h_1 - h_2]$ versus x/L, for various ratios of the outflow head to the inflow head, h_2/h_1, where h_2 ranges from 5% of h_1 to 95% of h_1. Note that curvature of $h(x)$ is most dramatic when the ratio h_2/h_1 is small. When this ratio is on the order of 80% or larger, the profile of $h(x)$ with distance is practically linear.

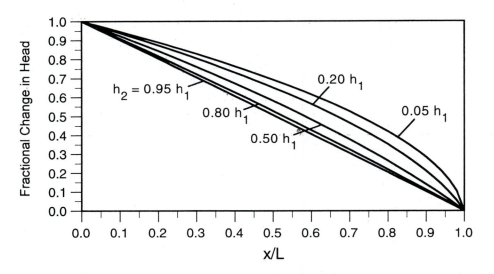

Figure 8.9 Example of $h(x)$ for 1-D unconfined flow under the Dupuit assumptions showing the normalized parameter $[h(x) - h_2]/[h_1 - h_2]$ versus x/L, for various ratios of h_2/h_1.

The corresponding total discharge. The corresponding total discharge is found by employing Darcy's law in the form

$$Q_x = -\frac{\partial \Phi}{\partial x} \qquad (8.61)$$

according to which we differentiate (8.57) with respect to x to yield the simple appearing result

$$Q_x = -\partial\Phi(x)/\partial x = -[(\Phi_2 - \Phi_1)/L] \qquad (8.62)$$

We next convert this to a more useful expression in terms of h. Upon substituting $\Phi = (Kh^2)/2$, we obtain the total discharge in terms of the hydraulic heads at the respective boundaries:

$$Q_x = -K\,[(h_2^2 - h_1^2)/2L] \qquad (8.63)$$

- 102 -

As we might have expected from the conservation of mass flux, Q_x is constant along the length of the aquifer. Its functional form should be no surprise, since it decomposes to

$$Q_x = - [K(h_2 + h_1)/2] [(h_2 - h_1)/L] \qquad (8.64)$$

The leading term can be looked upon as the *average transmissivity*

$$<T> = K(h_2 + h_1)/2 \qquad (8.65)$$

of an aquifer having an average hydraulic conductivity of K and an average thickness given by the mean of the elevations of the water table at either end, corresponding to $<h> = (h_2 + h_1)/2$.

The second term in square brackets in (8.64) is simply the average gradient of the hydraulic head

$$<\partial h/\partial x> = [(h_2 - h_1)/L] \qquad (8.66)$$

The corresponding specific discharge. To determine q_x in the saturated section of the aquifer (in the region $0 < z < h$), we use Darcy's law

$$q_x = - K \, \partial h /\partial x \qquad (8.67)$$

and differentiate $h(x)$ with respect to x, to determine the specific discharge

$$q_x = - K \{[(h_2^2 - h_1^2)/L] x + h_1^2\}^{-1/2} [(h_2^2 - h_1^2)/2L] \qquad (8.68)$$

While (8.68) appears to be somewhat complicated at first glance, upon closer inspection its form is not unexpected. It is, in fact, quite intuitive. Recall that $-K$ times the last term in square brackets in (8.68) is the total discharge

$$Q_x = - K [(h_2^2 - h_1^2)/2L] \qquad (8.69)$$

In addition, recall that this flow is assumed to be distributed uniformly over the thickness of the saturated layer or the local elevation of the water table given above by (8.60)

$$h(x) = \sqrt{h_1^2 + \frac{(h_2^2 - h_1^2)}{L} x} \qquad (8.60)$$

Note that $h(x)$ in the latter expression is precisely the denominator in (8.68), so that (8.68) can be represented by the quotient of (8.69) divided by (8.60), or

$$q_x(x) = Q_x/h(x) \qquad (8.70)$$

This expression simply states that the specific discharge q_x at a point along the x axis is given by the total discharge at that point divided by the saturated thickness of the aquifer. Not at all surprising.

From (8.70) we see that, since the discharge Q_x is constant, it should be clear that for the case illustrated in Figure 8.8 (as shown with $h_1 > h_2$), as x increases, the hydraulic head h *decreases* so that the flow per unit cross-sectional area intensifies, and the specific discharge q_x correspondingly *increases*.

Flow with Local Sources

Basic relations. We derived the following 3-D divergence relation in Chapter 5 for the specific discharge

$$\nabla \cdot \mathbf{q} = w_s \qquad (5.34)$$

where w_s represents the local production of water in units of the volume of water produced per unit volume of aquifer per unit time [or $L^3 L^{-3} T^{-1}$ for a composite dimension of T^{-1}]. We invoke the Dupuit assumptions and neglect any vertical component of flux associated with unconfined flow. Thus, if there are local sources of water within the aquifer illustrated in Figure 8.8, for 1-D flow in the x direction, (5.34) reduces to

$$dq_x/dx = w_s \qquad (8.71)$$

Assuming the production of water over the thickness of the unconfined aquifer is uniform, we can write the depth-integrated source term as

$$W_s = h w_s \qquad (8.72)$$

so that the divergence condition for the total discharge for source-free conditions given by (8.35) *now* becomes

$$dQ_x/dx = W_s \qquad (8.73)$$

Recalling Darcy's law for the discharge potential given previously by

$$Q_x = -\frac{\partial \Phi}{\partial x} \qquad (8.42)$$

we can substitute the right hand side of (8.42) for Q_x in (8.73) to obtain the following Poisson-type inhomogeneous differential equation:

$$d^2\Phi/dx^2 = -W_s \qquad (8.74)$$

In the following discussion, we want to relate the fluid production term, W_s, to the infiltration of water from precipitation. To do so we assume rain falls at the average rate of r_p meters per unit time. Thus water flows vertically through a unit horizontal area (e.g. 1 m^2) of the earth's surface at the rate of $q_z = r_p \times$ Unit Area [having dimensions of volume units per unit area per unit time]. We assume that this flux is instantaneously distributed over the thickness of the saturated section h, and is available for horizontal flow (c.f. the discussion related to Figure 4.4). Thus W_s in (8.74) can be directly equated to the rate of precipitation, or more precisely, the rate of infiltration.

Applying boundary conditions. Let us begin by assuming that the water channels in Figure 8.8 have the same hydraulic head, which we will set equal to the discharge potential Φ_0 associated with the static head h_0. As illustrated in Figure 8.10, the boundary conditions on the solution of our inhomogeneous differential equation (8.74) become simply

$$\Phi(0) = \Phi_0 \qquad (8.75a)$$

and

$$\Phi(L) = \Phi_0 \qquad (8.75b)$$

Integrating (8.74) twice with respect to x, we obtain the quadratic form

$$\Phi(x) = A + Bx - \frac{W_s}{2}x^2 \qquad (8.76)$$

Upon applying the boundary condition (8.75a) to (8.76), we obtain

$$A = \Phi_0 \qquad (8.77)$$

Next, applying boundary condition (8.75b) with (8.77) to (8.76), leads to

$$B = W_s/2 \qquad (8.78)$$

Thus the general form for the discharge potential given by (8.76) becomes

$$\Phi(x) = \Phi_o + \frac{W_s L}{2}x - \frac{W_s}{2}x^2 \qquad (8.79)$$

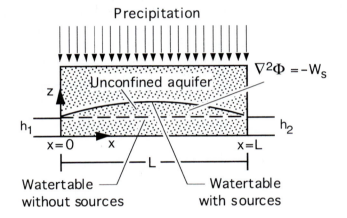

Figure 8.10 1-D unconfined aquifer with local infiltration. The hydraulic heads at x = 0 and x = L are held at the same unperturbed values: $h_1 = h_2 = h_o$.

With a little algebraic manipulation (8.79) can be rewritten as

$$\Phi(x) = \Phi_o + \frac{W_s}{2}\left(\frac{L}{2}\right)^2 - \frac{W_s}{2}\left(x - \frac{L}{2}\right)^2 \qquad (8.80)$$

The latter form is convenient in that it underscores the fact that $\Phi(x)$ is symmetric about, and has its maximum at, the mid-point $x = +L/2$.

Moreover, a simple differentiation with respect to x indicates, according to Darcy's law (8.42), that at x = 0,

$$Q_x = -\frac{W_s L}{2} \qquad (8.81)$$

which states that the total discharge is directed in the negative x direction, and is equal to one-half of the total precipitation falling on the aquifer. A similar operation at x = L results in

$$Q_x = + \frac{W_s L}{2} \qquad (8.82)$$

which accounts for the other half of the total precipitation falling on the aquifer, and which flows out of the system in the positive x direction.

Since the discharge potential is related to the hydraulic head through the expression (8.41), the variation of hydraulic head with distance x is given by

$$h(x) = \sqrt{ h_o{}^2 + \frac{W_s}{K} \left(\frac{L}{2}\right)^2 - \frac{W_s}{K} \left(x - \frac{L}{2} \right)^2 } \qquad (8.83)$$

Figure 8.11 shows the results of employing (8.83) to calculate the local elevation of the watertable $[h(x) - h_o]$ for various rates of annual infiltration, W_s. The hydrogeologic parameters are generally representative of the upper glacial aquifer for Long Island, New York. The width L = 30 km is approximately the distance between Long Island Sound and the Atlantic Ocean. The value of hydraulic conductivity K = 80 m d^{-1} is high, but typical for sand and gravel aquifers comprised of well sorted, stratified deposits. We assume that h_o is the thickness of the saturated section at the coast, approximately 60 m (200 ft). Various scenarios are plotted in Figure 8.11 for infiltration rates from 0.2 to 1.0 m yr^{-1} (meters per year).

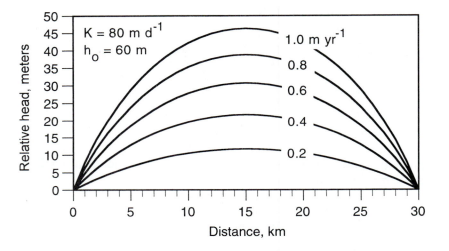

Figure 8.11 A 1-D unconfined flow model showing the local elevation of the watertable, $h(x) - h_o$, for various rates of annual infiltration, W_s. The hydrogeologic parameters are representative of the upper glacial aquifer for Long Island, New York.

The actual watertable over the interior of Long Island is raised to a maximum elevation of approximately 29 m (95 ft). From Figure 8.11 one notes that such a head increase would correspond to an annual infiltration rate of somewhat less than 0.6 m yr^{-1}.

Since the annual rainfall is approximately 1.1 m (44 in), this would imply that approximately 50% of the precipitation infiltrates to the watertable, with the remaining 50% apparently lost through evapotranspiration, overland flow and streamflow; a result that is consistent with more refined hydrologic analyses of this area.

Direct Integration of Regional Hydraulic Gradients with Local Sources

Flow relation. In the first part of this section, we considered the source-free behavior of the hydraulic head, then considered the case of local sources of water production, such as precipitation. We now want to combine these two cases, and analyze the situation in Figure 8.8 along with the source term effects illustrated in Figure 8.10. To do so, we directly integrate the inhomogeneous form (8.74) with the boundary conditions from our earlier discussion, which at $x = 0$ is

$$\Phi = \Phi_1 = (Kh_1^2)/2 \tag{8.46}$$

and at $x = L$ is

$$\Phi = \Phi_2 = (Kh_2^2)/2 \tag{8.47}$$

The general integral of (8.74) was given above as

$$\Phi(x) = A + Bx - \frac{W_s}{2}x^2 \tag{8.76}$$

Applying condition (8.46) to (8.76) at $x = 0$ leads to

$$A = (Kh_1^2)/2 \tag{8.84}$$

Applying condition (8.47) to (8.76) at $x = L$ leads to

$$\Phi(L) = \frac{Kh_2^2}{2} = \frac{Kh_1^2}{2} + BL - \frac{W_s}{2}L^2 \tag{8.85}$$

which can be solved for

$$B = \frac{K(h_2^2 - h_1^2)}{2L} + \frac{W_sL}{2} \tag{8.86}$$

Substituting for A and B in (8.76) we obtain

$$\Phi(x) = \frac{Kh_1^2}{2} + \frac{K(h_2^2 - h_1^2)}{2L}x + \frac{W_sL}{2}x - \frac{W_s}{2}x^2 \tag{8.87}$$

Upon invoking the definition of the discharge potential for unconfined flow given by (8.41) we can rewrite (8.87) in terms of the hydraulic head

$$h(x) = \sqrt{h_1^2 + \frac{(h_2^2 - h_1^2)}{L}x + \frac{W_s}{K}\left(\frac{L}{2}\right)^2 - \frac{W_s}{K}\left(x - \frac{L}{2}\right)^2} \tag{8.88}$$

Example. Figure 8.12 illustrates the results of applying (8.88) to a qualitative analysis of a field example from the Northeast Atlantic coastal plain of the United States. A glacial moraine parallel to the coast has deflected the regional drainage system such that local streams and lakes on the inland side of the moraine, a few kilometers from the ocean, are some 10 meters above sea level. Over a distance of approximately 2 kilometers, between the inland drainage system and the coast, precipitation has created a local groundwater divide having an elevation of between 20 to 25 meters.

We assume for the purpose of this calculation that the unconsolidated section extends to a depth of approximately 10 meters below sea level. Thus, for the present case, we assume $h_1 = 20$ m, $h_2 = 10$ m, $L = 2000$ m, and $W_s = 0.5$ m/yr (the latter being somewhat less than 50% of the annual precipitation). These parameters are employed in expression (8.88) to solve for the elevation of the watertable relative to sea level ($h_2 = 10$ m) for various assigned values of the hydraulic conductivity indicated in Figure 8.12.

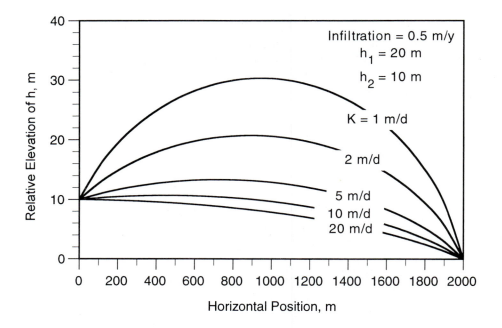

Figure 8.12 Results for a 1-D unconfined flow model with infiltration showing the local elevation of the watertable relative to sea level. The hydrogeologic parameters are representative of the glaciated coastal plain of the Northeast United States.

While the model geometry clearly oversimplifies the local geology, it appears that a hydraulic conductivity of 2 m/d (± a factor of 2) provides a reasonable fit to the observation that infiltration has caused an induced elevation of the watertable beneath the moraine of between 20 to 25 m above sea level. That these glacial moraine deposits ($K = 2$ m/d) are significantly less conductive than the well sorted, stratified glacial deposits ($K = 80$ m/d) used for the example in Figure 8.11 is inferred to be due to the heterogeneous distribution of grain sizes associated with poorly sorted moraine deposits.

Chapter 9. Natural Steady State Recharge and Discharge Systems in the Vertical Plane

GENERAL STATEMENT OF THE PROBLEM

Two-Dimensional Flow in the Vertical Plane

In this chapter we consider a relatively idealized model representing topographically driven subsurface flow on a regional scale. We assume that precipitation recharges the groundwater system such that fluid flows from higher elevations (or more precisely from areas of higher hydraulic potential) to lower elevations where it discharges to a laterally flowing stream or river that, in vertical section, carries excess fluid out of the plane of the page.

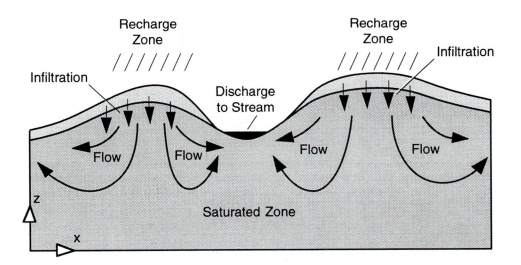

Figure 9.1 Conceptual model of a recharge-discharge system in vertical section.

We assume that subsurface flow is restricted to the x-z plane, so that the following parameterizations apply. The specific discharge vector in two dimensions is given by

$$\mathbf{q} = q_x \hat{\mathbf{x}} + q_z \hat{\mathbf{z}} \qquad (9.1)$$

Darcy's law in vector form

$$\mathbf{q} = - K \,\text{grad}\, h = - K \nabla h \qquad (9.2)$$

in 2-D is

$$\mathbf{q} = -K \frac{\partial h}{\partial x} \hat{\mathbf{x}} - K \frac{\partial h}{\partial z} \hat{\mathbf{z}} \qquad (9.3)$$

with the components

$$q_x = -K \frac{\partial h}{\partial x} \qquad (9.4)$$

- 109 -

and

$$q_z = -K \frac{\partial h}{\partial z}$$ (9.5)

Within the modeling region, we assume that the divergence of flux is source-free, so that the divergence relation is

$$\text{div } \mathbf{q} = \nabla \cdot \mathbf{q} = 0$$ (9.6)

which in 2-D becomes

$$\frac{\partial q_x}{\partial x} + \frac{\partial q_z}{\partial z} = 0$$ (9.7)

Upon substituting (9.4) and (9.5), respectively, for q_x and q_z in the divergence relation (9.7), we obtain the following form for Laplace's equation

$$\frac{\partial^2 h}{\partial x^2} + \frac{\partial^2 h}{\partial z^2} = 0$$ (9.8)

Our objective is to seek solutions of Laplace's equation that are compatible with the boundary conditions appropriate to the specific problem at hand. Having obtained $h(x,z)$ throughout our modeling region, we can then plot the equipotential surfaces, or in 2-D the h-lines

$$h(x,z) = h_{ci}$$ (9.9)

where h_{ci} represents the constant value assigned to the ith h-line.

We can next differentiate the function $h(x,z)$ using (9.4) and (9.5) to obtain the respective vector components of specific discharge, q_x and q_z.

In addition, we adapt our discussion on flow nets in Chapter 6 to express the total flux flowing between two points in the vertical plane, $P_o = (x_0,z_0)$ and $P_1 = (x_1,z_1)$, in the form

$$Q_{2D} = \int_{(x_0,z_0)}^{(x_1,z_1)} \mathbf{q} \cdot \hat{\mathbf{n}} \, dl$$ (9.10)

where several representative integration paths are shown in Figure 9.2.

Note that, to be consistent with our convention in Chapter 6 on flow nets, the unit normal vector is given by

$$\hat{\mathbf{n}} = - \, d\mathbf{l} \times \hat{\mathbf{y}} \, / \, |d\mathbf{l}|$$ (9.11)

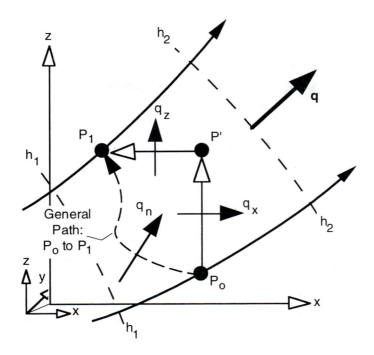

Figure 9.2 Convention for defining Q_{2D} for integration paths in the x-z plane.

With this convention we can then integrate

$$V(x,z) = V_o + \int_{(x_o,z_o)}^{(x,z)} [\, q_x dz - q_z dx \,] \qquad (9.12)$$

to obtain the flow function $V(x,z)$, from which we can determine the corresponding flow-lines

$$V(x,z) = V_{cj} \qquad (9.13)$$

where V_{cj} is the constant assigned to the jth flow-line.

A SPECIFIC SOLUTION

A Simple Harmonic Model for the Potentiometric Surface

In the spirit of keeping the analysis as basic as possible, while still assuring a non-trivial result, we will consider a simplified adaptation of the elegant approach of Toth (1963). The topography of the potentiometric surface will be represented by a constant elevation value, h_o, on which we superimpose a simple trigonometric function

$$\Delta h(x) = A \cos\left(\frac{2\pi(x - x_o)}{\lambda_s} \right) \qquad (9.14)$$

having a maximum amplitude A that occurs at $x = x_0$, and a characteristic spatial wavelength λ_s. For the present purpose we will define

$$\lambda_s = 2\,L_{RD} \qquad (9.15)$$

where L_{RD} is the characteristic distance between the recharge and discharge systems. For the case shown in Figure 9.1, where we want to consider discharge to a stream at the center of our model ($x = 0$), we write $h(x)$ in the form

$$h(x) = h_0 - A\cos\left(\frac{2\pi(x - x_0)}{\lambda_s}\right) \qquad (9.16)$$

where, to be specific, $h_0 = 100$ meters, $\lambda_s = 100$ meters, $A = 10$ meters, and $x_0 = 0$. Thus, as shown in Figure 9.3, the potentiometric surface is at an elevation of 110 meters beneath the recharge zone in the "hills" and at 90 meters elevation for the discharge zone in the "valley". Our objective is to study the nature of fluid circulation in a vertical plane transecting these two systems.

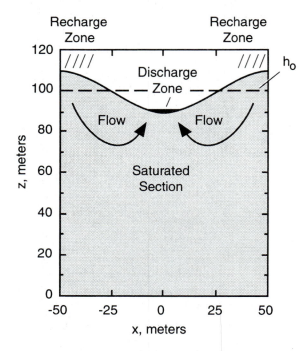
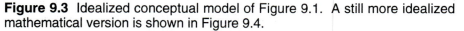

Figure 9.3 Idealized conceptual model of Figure 9.1. A still more idealized mathematical version is shown in Figure 9.4.

Assumed and Implied Boundary Conditions

We assume that our modeling region spans a range in x

$$-\,L_{RD} \leq x \leq +\,L_{RD} \qquad (9.17)$$

and a region in z

$$-\,\infty \leq z \leq +\,h_0 \qquad (9.18)$$

Within the interior of this region there are no sources. A fundamental boundary condition is that the lower boundary of the aquifer is sufficiently deep (- ∞) to have negligible effects on our results. This is because, at this stage of our analysis, we want only to explore the coupling between the recharge and discharge zones, and how this coupling affects the deep circulation of fluid. We do not want to address the interaction of this flow with the lower boundary.

A second fundamental boundary condition is that at the top boundary, at a specified elevation $z = h_o$, we will assume that conditions of recharge and discharge are naturally adjusted so that a small perturbation in hydraulic head is applied of the form (9.16). This becomes the boundary condition on the potentiometric function applied at $z = h_o$ along the top surface of our region.

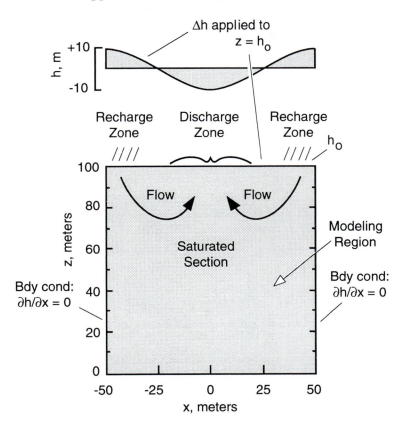

Figure 9.4. The rectangular model space for the problem considered in the text. A cosine-like forcing function is applied at $z = h_o$. The lower boundary is assumed to be at $z = -\infty$.

As consequences of the latter conditions on the hydraulic head, the following "boundary" conditions are implied. The symmetry of the problem dictates that $\partial h/\partial x = 0$ at the left and right vertical boundaries, so that there is no net flow out of the modeling region. Thus with reference to Figure 9.2, we have

$$\left. \frac{\partial h}{\partial x} \right|_{\substack{\text{Vert} \\ \text{Bdy}}} \equiv 0 \qquad (9.19)$$

along the two vertical boundaries at

$$x = \pm L_{RD} \qquad (9.20)$$

Since the hydraulic head is concave down at $x = \pm L_{RD}$, precisely beneath the peak elevations of the recharge zones, subsurface flow must be strictly vertically downward. We also note that a vertical plane of symmetry exists for h precisely through the lowest elevation in the trough of the discharge zone at $x = 0$. This also implies that there is no lateral flow in the x direction across the vertical plane at $x = 0$; but since this symmetry is associated wth a *minimum* in the hydraulic head (from 9.16), subsurface flow must be strictly vertically *upward*. For either of these two kinds of vertical planes of symmetry — recharge or discharge — there is no lateral flow in the x direction across these vertical planes.

Solving Laplace's Equation by Separation of Variables

Following the time-proven technique from applied mathematics known as the *method of separation of variables*, we assume that a function of x and z that satisfies Laplace's equation can be written as the product of two functions, one of x and one of z, so that

$$h(x,z) = X(x)Z(z) \tag{9.21}$$

where $X(x)$ is a function only of x, and $Z(z)$ is a function only of z.

That (9.21) is in fact a solution to Laplace's equation can be seen by substituting (9.21) into (9.8). Upon differentiating

$$Z(z) \frac{\partial^2 X(x)}{\partial x^2} + X(x) \frac{\partial^2 Z(z)}{\partial z^2} = 0 \tag{9.22}$$

which can be rearranged to the form

$$\frac{1}{X(x)} \frac{\partial^2 X(x)}{\partial x^2} = - \frac{1}{Z(z)} \frac{\partial^2 Z(z)}{\partial z^2} \tag{9.23}$$

Thus (9.21) is a solution to Laplace's equation providing the equality holds. The left hand side of (9.23) is a function only of x, and the right side is a function only of z. The only way that the two expressions can be equal is that they are equal to the same constant, so that we obtain the two ordinary differential equations

$$\frac{1}{X(x)} \frac{d^2 X(x)}{dx^2} = C \tag{9.24}$$

$$\frac{1}{Z(z)} \frac{d^2 Z(z)}{dz^2} = -C \tag{9.25}$$

which can be rearranged to the forms

$$\frac{d^2 X(x)}{dx^2} = C\,X(x) \tag{9.26}$$

$$\frac{d^2Z(z)}{dz^2} = -C\,Z(z) \qquad (9.27)$$

We can see that a solution to (9.26) is a simple harmonic function, such as a sine or cosine function of x, through a simple substitution of

$$X(x) = Const_1 \cdot \cos(kx) + Const_2 \qquad (9.28)$$

or

$$X(x) = Const_3 \cdot \sin(kx) + Const_4 \qquad (9.29)$$

where the wave number k is given by

$$k = \frac{2\pi}{\lambda} \qquad (9.30)$$

λ being the horizontal wavelength. For example, substituting (9.28) into the left hand side of (9.26) and differentiating twice, leads to

$$-k^2 \cos(kx) = C \cos(kx) \qquad (9.31)$$

Thus (9.28) is a solution to (9.26) providing the separation constant is given by

$$C = -k^2 \qquad (9.32)$$

In the jargon of applied mathematics, we say that (9.28) is an "eigenfunction" solution to the ordinary second order differential equation (9.26), providing the "eigenvalue relation" (9.32) holds. In this case, k is termed an "eigenvalue".

A similar result would follow if the sine function given by (9.29) were chosen. Which, if either or both, of these functions is appropriate depends on the boundary conditions. Since the boundary condition at $z = h_o$ calls for a cosine type function centered at $x = 0$, we will accept (9.28) as the solution for the $X(x)$ term.

Since the separation constant C is now required to have the special form (9.32), equation (9.27) for Z(z) takes on the special form

$$\frac{d^2Z(z)}{dz^2} = k^2\,Z(z) \qquad (9.33)$$

Possible solutions to (9.33) are given by

$$Z(z) = Const_5 \cdot e^{+kz} \qquad (9.34)$$

or

$$Z(z) = Const_6 \cdot e^{-kz} \qquad (9.35)$$

either of which can be verified by substitution into (9.33).

Note: While the "+" sign in the argument of the exponential term in (9.34), strictly speaking, is superfluous, we include it here and in the following discussion, however, to explicitly emphasize that we are dealing with the positive root of k^2.

Combining (9.28), (9.29), (9.34) and (9.35), into the solution originally posed by (9.21), the complete solution for h(x,y) becomes

$$h(x,y) = A + [B \cos (kx) + C \sin (kx)] e^{+kz}$$
$$+ [D \cos (kx) + E \sin (kx)] e^{-kz} \qquad (9.36)$$

where the respective constants A, B, C, D and E will be determined by the details of the boundary value problem being considered.

Casting the constants in terms of the present problem, we first note that the forcing term for the hydraulic head given by (9.16) has only a cosine dependence, so the sine terms in (9.36) are not needed, and we can set

$$C = E = 0 \qquad (9.37)$$

Therefore (9.36) for the particular case considered here reduces to

$$h(x,y) = A + B \cos (kx) e^{+kz} + D \cos (kx) e^{-kz} \qquad (9.38)$$

We now note that exponential terms with intrinsically positive arguments *decrease* with decreasing z, and terms with intrinsically negative arguments *increase* with decreasing z. Because the last term on the right side of (9.38) would increase without limit with increasing depth (in the -z direction), it is therefore not consistent with the present boundary conditions, and we set the constant coefficient

$$D = 0 \qquad (9.39)$$

Note: This latter term, however, is essential if we wanted to explore the effects of an impermeable boundary at the base of the aquifer at some finite depth, or layering within the aquifer. But this is beyond the present treatment.

Accordingly, the solution to our present problem is

$$h(x,z) = A + B \cos (kx) e^{+kz} \qquad (9.40)$$

which, upon adjusting the various constants to represent the parameters defined above, becomes

$$h(x,z) = h_o - \Delta h \cos\left(\frac{\pi x}{L_{RD}}\right) e^{+k(z - h_o)} \qquad (9.41)$$

The exponential term has been adjusted so that, at the top boundary of the modeling region (which is set at $z = h_o$), its argument is zero, and the exponential function itself is unity. Thus (9.41) reduces to the potentiometric forcing term (9.16) at $z = h_o$.

One should note that, while the remaining exponential term in (9.41) has the rather non-physical property of growing without bound for $x > h_o$, this region is excluded from our model space. It is emphasized that for the particular model considered here, the solution is only valid for the rectangular model space shown in Figure 9.4. This is a common aspect of most such mathematical models: one determines a proper solution over their defined range of model space; but such a solution may be totally meaningless outside that range.

Application to a Theoretical Example

The results of such an analytical model are shown in Figure 9.5, where an equidimensional 100 x 100 meter region of the model space is shown. The half-wavelength distance between recharge and discharge zones is 50 meters, and the distance between recharge zones is accordingly 100 meters.

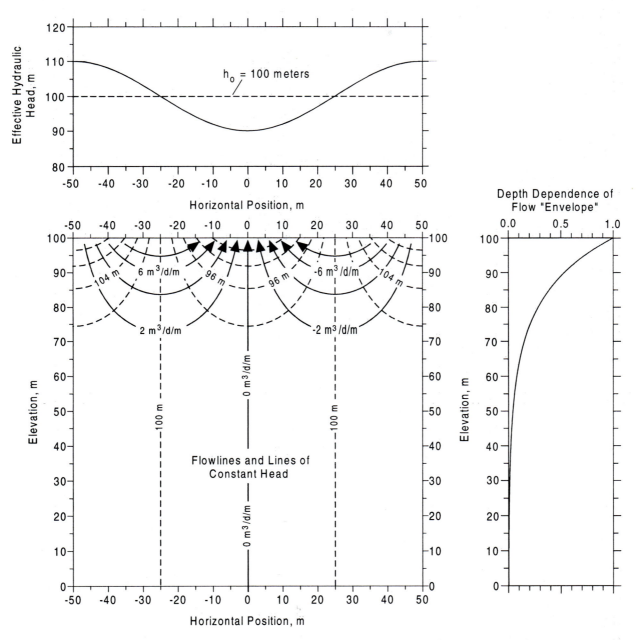

Figure 9.5 Flow-lines and potential lines for an analytical model. A cosine-like variation in hydraulic head of ±10 meters is superimposed on a static elevation of 100 meters. The distance between recharge zones is 100 meters. The exponential envelope for exp(kz) is shown in the right-most panel.

A variation in hydraulic head of ±10 meters is superimposed on a static elevation of 100 meters. The condition that there is no horizontal flow on the left and right sides of the area shown follows from the symmetry of h applied at the top surface. While the model space extends to infinite depth, it is clear from the contours of the flow-lines that fluid circulation is restricted to the uppermost section. This is due to the exponential decrease of the potential functions with depth, as seen in the right-most panel, where the exponential envelope for $\exp(kz)$ is shown for $k = 2\pi / 100$, and z becomes increasingly more negative downward. Thus, while in principle water can circulate to infinite depth since the exponential factor remains finite for $z < -\infty$, in fact flow rates are significant only in a zone extending to a depth substantially less than a spatial wavelength, $\lambda_s = 2 L_{RD}$, where as defined above L_{RD} is the characteristic distance between the recharge and discharge systems.

A reasonable measure of the *depth of circulation* is when the exponential term decreases to $1/e$th of its surface value, so that significant flow rates are restricted to the zone: $d \leq 1/k = \lambda_s/2\pi$. Conversely, if the lateral distance between recharge and discharge zones is large, then circulation can occur to great depth.

Chapter 10. Steady State Flow to a Well

GENERAL

The mathematical analysis of subsurface flow to a well discharging at the surface is a classic problem in hydrogeology, not only because it provides a basis for assessing the qualities of an aquifer, but also because the expressions for discrete point sources or point sinks can be superimposed, or in a sense integrated, to represent more complex natural or manmade distributed systems. To many workers, the expression for flow to a point sink (a discharging well) is as fundamental to hydrology as is Newton's law of gravitation to astronomy. In this chapter, after a brief discussion of the transition from transient to equilibrium flow from a discharging well, we will emphasize the analysis of the equilibrium — or steady-state — case; and derive the so-called *Thiem* relation for steady-state subsurface flow. In the next chapters we will consider transient (i.e. non-equilibrium) flow in some detail.

TRANSITION FROM TRANSIENT TO STEADY-STATE FLOW AT A DISCHARGING WELL

Drawdown of h and the "Cone of Depression"

If fluid is extracted *from* the aquifer, we term this a "discharge", and if fluid is injected *into* the aquifer, we term this a "recharge". Here, to be specific, we consider radial flow to a well discharging at a constant flow rate Q [units: m^3/s; dimensions: L^3/T].

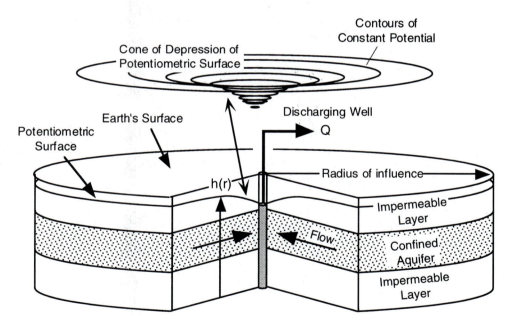

Figure 10.1. Radial flow to a well discharging from a confined aquifer.

If fluid flows *towards* a well, from Darcy's law there must be a hydraulic gradient such that the hydraulic head h decreases as one approaches the well. Since flow is assumed to be symmetric about the well, such a flow pattern creates what is referred to as a "cone of depression" of the potentiometric surface, as shown in Figure 10.1.

This figure shows a confined aquifer, that is to say, a relatively permeable layer sandwiched between two impermeable layers. Before any pumping occurs, it is assumed that the water was at some uniform *static head* h_o. For a confined aquifer, this would correspond to a horizontal potentiometric surface. For an unconfined aquifer, this would correspond to a horizontal water table or free surface. Local gradients of the potentiometric or groundwater surface may, and usually do, exist before pumping begins, but their effects are not considered here.

This chapter deals largely with representing *steady-state* flow; namely what has become known as the Thiem solution to the problem. But to understand the implications of the underlying assumptions of the steady-state analysis, we will first consider some general aspects of *transient* flow.

Transient Conditions under Constant Discharge

Figure 10.2 illustrates the hydraulic head, $h(r,t)$, in the vicinity of a well discharging at a constant rate from a confined aquifer having a transmissivity (depth-integrated conductivity) of $T = 0.1$ ft^2 min^{-1} (13.4 m^2 day^{-1}) and a storativity of $S = 0.00025$ (defined in Chapter 11). (We employ English units here in deference to common practice among community and regulatory agencies.)

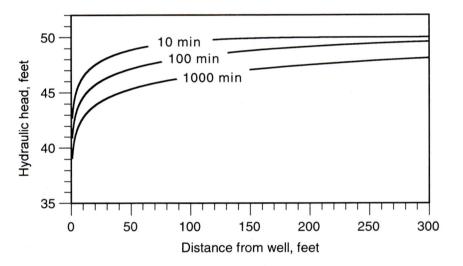

Figure 10.2 Examples of the drawdown of hydraulic head (the so-called "cone of depression") in the vicinity of a well discharging from a confined aquifer at three instants of time: 10, 100, and 1000 minutes.

Assume that at $t = 0$, our test well begins discharging at a constant flow rate of $Q = 1.0$ ft^3/min; the equivalent of 7.48 gallons per minute; or in SI units: 0.472×10^{-3} m^3 s^{-1}, 0.472 liter s^{-1}, or 40.8 m^3 day^{-1}. This is a rather large value for a single family home, but modest for a community supply. Drawdown is shown in Figure 10.2 as a function of distance for three instants of time — what might be termed from "early time" to "late time".

The instant pumping begins, a transient condition is imposed on the aquifer, and a rapid drawdown of water occurs in the immediate vicinity of the well. The so-called "cone of depression" (a cusp in the 2-D vertical section) at early pumping time is very steep and tightly focussed on the discharging well. As shown in Figure 10.2, as time progresses, the cone deepens, and expands to a larger radius of influence. With reference to the confined aquifer in Figure 10.1, the cone of depression, at a particular snapshot in time, would be observed as a progressive decrease in the elevation of the water level in monitoring wells (an increase in the depth below grade of the potentiometric surface) as the discharging well is approached. For unconfined aquifers, one would actually see a lowering of the water table near the discharging well. The decrease in h relative to its pre-pumping static level h_o, whether for confined or unconfined aquifers, is called "*drawdown*", and is defined here as $h_o - h$. Note that, according to this convention, *drawdown* in the vicinity of a discharging well has intrinsically *positive* values.

Comparing Transient Drawdown in Two Monitoring Wells

Continuing to employ the example of the last section and Figure 10.2, we now compare the drawdown in two monitoring wells installed in the vicinity of the discharging well in order to assess the characteristics of the aquifer. One well is located at $r_1 = 50$ ft (15.2 m) from the discharging well, the other at $r_2 = 100$ ft (30.4 m). As above, we assume that at t = 0, our test well begins discharging at a constant flow rate of Q = 1.0 ft^3 min^{-1} (0.472 x 10^{-3} m^3 s^{-1}, or 40.8 m^3 day^{-1}). Figure 10.3 summarizes the drawdown in each well from t = 0 to t = 5000 minutes (3 x 10^5 s), somewhat longer than 3 days.

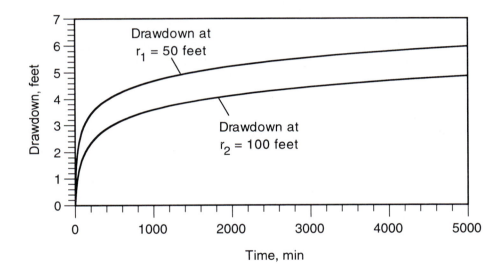

Figure 10.3 The drawdown with time in two monitoring wells at distances of $r_1 = 50$ ft and $r_2 = 100$ ft from the discharging well. Conditions are the same as Figure 10.2.

As an aside, we should note that such a pumping period is very long for a domestic well, but not atypical for a community well. The issue in such an extended test is where to run the discharging fluid so that it does not reenter the system, and "short circuit" the drawdown of the actual aquifer.

With reference to the example in Figure 10.3, after 3 days of pumping, it is clear that there continues to be a measurable increase in drawdown with time. In fact, as we will see in a later chapter, the drawdown in our monitoring wells would increase continuously in time for the model assumed here. This has led more than one worker to question the relevance of the concept of "equilibrium" or "steady-state" flow for such well tests. There is in fact *little* relevance if one insists that the drawdowns, or the hydraulic heads themselves, achieve levels that are constant in time. Such is simply not consistent with the physics of flow, unless one pumps for a sufficiently long time on an island where one is completely surrounded by a recharge reservoir of infinite capacity (as envisaged by Todd, 1964; and Strack, 1989).

Some workers have pointed out that while, *theoretically,* drawdowns do not stabilize, *in practice* they often do from leakage through the less than perfect confining layers, recharge from nearby sources, and so forth (e.g. de Marsily, 1986). Thus, it is argued by some that since, to all intents and purposes, drawdowns do in fact stabilize to constant levels in many actual field tests for whatever reason, then a relatively simple steady-state solution to the flow equation — known as the above-mentioned *Thiem* solution, and discussed in some detail below — can be applied. Such is not the position of this text. While recognizing that in practice hydraulic heads in the vicinity of a discharging well often tend to stabilize after a finite time, this does not get around the issue of non-equilibrium drawdown for the ideal case considered above. To require stable or non time dependent hydraulic heads as a criterion to employ the Thiem relation is to invoke the wrong mathematical model to simulate a physical condition; it is to skirt the fundamental nuance of the Thiem assumptions, and the profound significance of their implications. As we see below, the Thiem method *does not* require steady-state drawdowns, only steady-state *flow* — but this is getting ahead of our discussion.

Defining "Equilibrium"

The fact is that one should not and, as we have seen, *cannot* invoke equilibrium conditions on the hydraulic heads. But, the beauty of the Thiem assumptions is that it is quite reasonable to invoke equilibrium or steady-state conditions on the *flow*. To see this, we need to consider how fluid flow could be monitored in situ in the aquifer in our thought experiment. The answer is obviously through some variation of Darcy's law where one measures the *difference* in hydraulic head between two monitoring wells, such as

$$\Delta h = h_2 - h_1 \qquad\qquad (10.1)$$

This might eventually be reduced to some type of quasi-gradient by dividing by the difference in their distance to the discharging well. But for our present purpose, we will simply analyze the parameter Δh as a function of time, where as shown in Figure 10.4 we have replotted the early

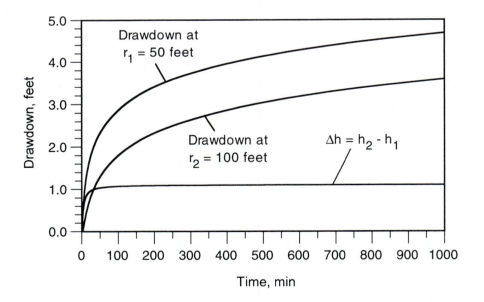

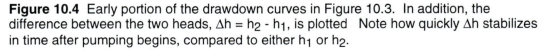

Figure 10.4 Early portion of the drawdown curves in Figure 10.3. In addition, the difference between the two heads, $\Delta h = h_2 - h_1$, is plotted Note how quickly Δh stabilizes in time after pumping begins, compared to either h_1 or h_2.

portion of the drawdown curves in Figure 10.3, with the addition of a curve representing their difference given by (10.1).

What is truly remarkable in these results is how quickly the difference $\Delta h = h_2 - h_1$ stabilizes in time, as opposed to either of the drawdowns Δh_1 or Δh_2, each of which progessively increases to ever larger values. Thus we conclude that the hydraulic *gradients* in transient flow may equilibrate very quickly in the vicinity of a discharging well, in spite of the fact that the hydraulic heads themselves never do. This in turn implies that the concept of equilibrium flow is quite reasonable, but that the concept of the hydraulic heads reaching equilibrium is not.

We define what we will term as the "radius of influence" as that radius within which transient flow has stabilized after a given time during transient drawdown. During an actual well test, one would record at various time intervals the drawdowns Δh_1 and Δh_2 (that is, the static level h_o minus the h in wells at two radii r_1 and r_2, respectively) and use the time when the difference in h stabilizes as a surrogate to determine when the flow has equilibrated.

STEADY-STATE FLOW TO A WELL DISCHARGING FROM A CONFINED AQUIFER

The Aquifer

We assume that pumping has proceeded for a sufficiently long time for equilibrium flow in the aquifer to have been established — this is what we define as "steady-state" recharge or discharge. Recall that the definition of a "confined aquifer" (also known as an "artesian aquifer" by some workers) is one that is bounded above and below by impermeable layers (termed "confining layers"), and whose fluid has a hydraulic head exceeding the elevation of the top of the aquifer. In operational terms, fluid in a confined aquifer is under sufficient pressure to force it to rise in a well bore above the top of the aquifer.

In this chapter we will be even more specific in the qualities of the confined aquifers that we will consider. We will assume that the aquifer is horizontal in the x-y plane, has uniform thickness b, and has a uniform hydraulic conductivity K. We will also assume that our well fully penetrates the aquifer, and that any recharge or discharge at a particular point P(x,y) takes place continuously and uniformly over its entire thickness. This allows us to assume uniform flow in the vertical plane, so that lines or planes of constant h (the equipotential surfaces) are strictly vertical, and perpendicular to the local flow direction.

Horizontal Flow and the Discharge Potential

Because it is so common to describe horizontal flow problems in terms of the total discharge over the integrated thickness b of the aquifer, we will make use of the two-dimensional discharge vector

$$Q = (Q_x, Q_y) \qquad (10.2)$$

where

$$Q_x = b \, q_x \qquad (10.3)$$

$$Q_y = b \, q_y \qquad (10.4)$$

and, of course, q_x and q_y are the vector components of specific discharge.

Darcy's law for the total discharge leads to the following two components

$$Q_x = b \, q_x = - b \, K \, \partial h / \partial x \qquad (10.5)$$

$$Q_y = b \, q_y = - b \, K \, \partial h / \partial y \qquad (10.6)$$

Following Strack (1989) we note that since b and K are each constant, these two expressions are equivalent to

$$Q_x = - \partial \, [bKh] \, / \partial x \qquad (10.7)$$

$$Q_y = - \partial \, [bKh] \, / \partial y \qquad (10.8)$$

and are motivated to define a new variable, Φ, which we will term the *discharge potential for horizontal confined flow*, having the form

$$\Phi(x,y) = b \, K h(x,y) + C_c \qquad (10.9)$$

where C_c is an arbitrary constant. Darcy's law relating Q to the gradient of Φ accordingly becomes

$$Q_x = - \partial \Phi / \partial x \qquad (10.10)$$

$$Q_y = - \partial \Phi / \partial y \qquad (10.11)$$

For horizontal flow, $\nabla \cdot \mathbf{q} = 0$ implies the following divergence relation for the discharge vector:

$$\frac{\partial Q_x}{\partial x} + \frac{\partial Q_y}{\partial y} = 0 \qquad (10.12)$$

Upon substituting the right sides of (10.10) and (10.11) from Darcy's law for Q_x and Q_y in the divergence relation (10.12), we obtain the following two-dimensional form for Laplace's equation

$$\frac{\partial^2 \Phi}{\partial x^2} + \frac{\partial^2 \Phi}{\partial y^2} = 0 \qquad (10.13)$$

Radial Flow to a Discharging Well

Assume that a well fully penetrates a confined aquifer having a uniform thickness of b, and an average hydraulic conductivity of K. The well pumps at a constant rate of Q_o, where if positive represents discharge of fluid *from* the subsurface, and if negative represents recharge *into* the subsurface. To be specific, we will assume for the following discussion that we are *discharging* from the well and that Q_o is implicitly positive. For this situation, flow is strictly horizontal and radial in the plane of the aquifer, and uniform over its thickness b.

Because of this symmetry, the hydraulic head will be radially symmetric, and for the steady-state case considered here, at a given instant of time, can be represented uniquely as a function only of the radius, which is to say as h(r). We now reassess the divergence relation for flow in an aquifer when it is directed radially to or from a well.

Conservation of Radial Flux

We begin by defining the coordinate parameters illustrated in Figure 10.5, where we assume there is a discharging (or recharging) well located at $r = 0$.

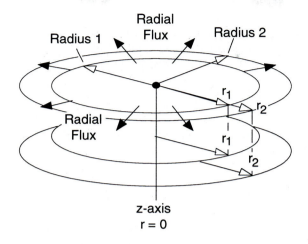

Figure 10.5 The cylindrical coordinate system, centered at a discharging (or recharging) well at $r = 0$. We want to analyze the flux at radii r_1 and r_2.

We apply the conservation condition

$$\oiint_{\text{Surface}} \mathbf{q} \cdot d\mathbf{A} = \iiint_{\text{Vol}} [\text{All sources}] \, dV \qquad (10.14)$$

to the annular volume depicted in Figure 10.6.

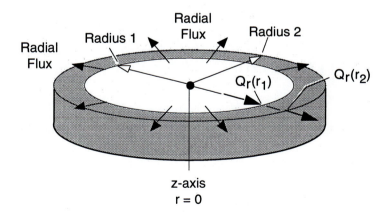

Figure 10.6 An annular volume (a thick circular ring, having a rectangular cross-section) to which we apply the conservation integral.

We assume that fluid flow has stabilized in time such that the net outward (positive) flux of material from this finite volume equals the rate at which material is produced by sources within the volume. To be specific, we express (10.14) in the form

$$\oiint_{\text{Surface}} \mathbf{q \cdot dA} = \iiint_{\text{Vol}} w(x,y,z) \, dV \qquad (10.15)$$

where $w(x,y,z)$ represents water production — or loss — per unit volume. Let us assume, for simplicity, that w is uniform (i.e. constant) throughout whatever volume we select. Moreover, we have assumed previously that all flow is radial and, at a given radius r, uniform over the thickness of the aquifer, such that

$$\mathbf{q} = q_r(r) \, \hat{\mathbf{r}} \qquad (10.16)$$

Thus there is no direct flow through the top or bottom of the annulus in Figure 10.6, so that the integral of normal flux for these two surfaces is zero.

Let r_1 be the inner radius and r_2 the outer radius of the annulus, and let its height be the thickness of the aquifer b. Both the left hand side and right hand side of (10.15) involve an integration over the vertical variable z which, because of the vertical uniformity of flow and the uniformity of w, simply becomes a multiplication by b. Thus the left hand side of (10.15) reduces to

$$\oiint_{\text{Surface}} \mathbf{q \cdot dA} = \left[q_r(r_2) \, b \int_0^{2\pi} r_2 \, d\theta \right] - \left[q_r(r_1) \, b \int_0^{2\pi} r_1 \, d\theta \right] \qquad (10.17)$$

Upon integrating over each circumference, and noting that the total discharge $Q_r = q_r \, b$ [dimensions: $L^2 T^{-1}$], (10.17) becomes

$$\oiint_{\text{Surface}} \mathbf{q \cdot dA} = \left[2\pi r_2 Q_r(r_2) \right] - \left[2\pi r_1 Q_r(r_1) \right] \qquad (10.18)$$

The right hand side of (10.15) becomes the uniform production (or loss) term w multiplied by the volume of the annulus, or

$$\iiint_{\text{Vol}} w(x,y,z) \, dV = w \left[b \, \pi \left(r_2^2 - r_1^2 \right) \right] \qquad (10.19)$$

Thus, the conservation condition (10.15), for the case of a finite-sized annulus, becomes

$$\left[2\pi r_2 Q_r(r_2) \right] - \left[2\pi r_1 Q_r(r_1) \right] = w \left[b \, \pi \left(r_2^2 - r_1^2 \right) \right] \qquad (10.20)$$

which upon cancelling common terms on either side reduces to

$$r_2 Q_r(r_2) - r_1 Q_r(r_1) = w \left[\frac{b}{2} \left(r_2^2 - r_1^2 \right) \right] \qquad (10.21)$$

Expression (10.21) becomes our conservation of flux condition for the specific case of the annular volume in Figure 10.6. Consider the units of this relation. The source (or sink) term w is parameterized in terms of the units of volume of fluid produced (or lost) per unit time per unit volume of the aquifer [dimensions: $L^3 T^{-1} L^{-3}$]. The units of the specific discharge q_r are length per time (such as m s^{-1}); and the units of the depth-integrated discharge $Q_r = q_r b$ are $m^2 s^{-1}$. Of course, each of the two terms on the left side of (10.21), and the single term on the right side must have the same units. Since we divided (10.20) by 2π radians to obtain the form (10.21), each of the three terms in the latter expression must then have dimensions of L^3 per unit time per radian, or total volume per unit time per radian (such as $m^3 s^{-1}$ radian^{-1}, or gallons per minute per radian, etc.).

Note: Although we preclude the contribution from *direct* flow through the top and bottom of the annulus on the left hand side of (10.15), we could implicitly allow for *indirect* vertical flow, such as infiltration or exfiltration through the confining layers by simulating its contribution as a "source" or "sink" on the right side of 10.15. Consider a source such as precipitation infiltrating from the surface at a rate p_r, given in terms of the depth of rainfall per unit time; or equivalently in terms of the volume of fluid per unit time crossing a unit area at the top of the aquifer [dimensions: $L T^{-1}$]. In such a case, we assume that this influx is instantaneously distributed over the thickness b of the aquifer, in effect causing a volumetric fluid production equivalent to $p_r = w b$ in (10.21).

Taylor's Series for Q_r

We can expand the total radial discharge Q_r as a function of the radius in the vicinity of point r_1

$$Q_r(r) = Q_r(r_1) + (r - r_1) \left. \frac{dQ_r}{dr} \right|_{r_1} + \frac{(r - r_1)^2}{2} \left. \frac{d^2 Q_r}{dr^2} \right|_{r_1} + \cdots \qquad (10.22)$$

We assume that r_1 is the inner radius of our annular volume, and $r = r_2$ is the outer radius. Note that all derivatives in (10.22) are evaluated at r_1. We also assume that the difference between the two radii

$$\Delta r = r_2 - r_1 \qquad (10.23)$$

is small, so that terms beyond the first order in (10.22) can be neglected. Thus

$$Q_r(r_2) = Q_r(r_1) + \Delta r \left. \frac{dQ_r}{dr} \right|_{r_1} \qquad (10.24)$$

Divergence of Radial flow. Substituting the right hand side of (10.24) for $Q_r(r_2)$ on the left hand side of (10.21), we have

$$r_2 \left[Q_r(r_1) + \Delta r \left. \frac{dQ_r}{dr} \right|_{r_1} \right] - r_1 Q_r(r_1) = w \left[\frac{b}{2} \left(r_2^2 - r_1^2 \right) \right] \qquad (10.25)$$

Upon substituting the following

$$r_2 = r_1 + \Delta r \qquad (10.26)$$

for r_2 in (10.25), leads to

$$\left(r_1 + \Delta r \right) \left[Q_r(r_1) + \Delta r \left. \frac{dQ_r}{dr} \right|_{r_1} \right] - r_1 Q_r(r_1)$$
$$= w \left[\frac{b}{2} \left(\left(r_1 + \Delta r \right)^2 - r_1^2 \right) \right] \tag{10.27}$$

Combining terms and neglecting contributions of order Δr^2, we obtain

$$\left[r_1 Q_r(r_1) + \Delta r\, Q_r(r_1) + r_1 \Delta r \left. \frac{dQ_r}{dr} \right|_{r_1} \right] - r_1 Q_r(r_1)$$
$$= w \left[\frac{b}{2} \left(\left(r_1^2 + 2 r_1 \Delta r \right) - r_1^2 \right) \right] \tag{10.28}$$

which, upon cancelling terms, simplifies to

$$\Delta r\, Q_r(r_1) + r_1 \Delta r \left. \frac{dQ_r}{dr} \right|_{r_1} = w b r_1 \Delta r \tag{10.29}$$

Upon dividing through by $r_1 \Delta r$ we obtain

$$\left. \frac{dQ_r}{dr} \right|_{r_1} + \frac{Q_r(r_1)}{r_1} = wb \tag{10.30}$$

Since our choice of r_1 was completely arbitrary, (10.30) holds for any r, leading to the divergence condition for radial flow in a confined aquifer

$$\frac{dQ_r}{dr} + \frac{Q_r(r)}{r} = wb \tag{10.31}$$

Poisson's and Laplace's equations for radial flow. Invoking Darcy's law for radial flow

$$Q_r = - T \frac{dh}{dr} \tag{10.32}$$

and substituting the right hand side of (10.32) for Q_r in (10.31), we obtain

$$T \frac{d}{dr} \left(\frac{dh}{dr} \right) + \frac{T}{r} \frac{dh}{dr} = - wb \tag{10.33}$$

which can be rearranged to the following form of Poisson's equation

$$\frac{d}{dr} \left(\frac{dh}{dr} \right) + \frac{1}{r} \frac{dh}{dr} = - \frac{wb}{T} \tag{10.34}$$

We define a discharge potential for radial flow, analogous to (10.9), such that

$$\Phi(r) \ = \ Kbh(r) + C_c \tag{10.35}$$

This allows us to rewrite (10.33) in terms of Φ as

$$\frac{d}{dr}\left(\frac{d\Phi}{dr}\right) + \frac{1}{r}\frac{d\Phi}{dr} \ = \ - \ wb \tag{10.36}$$

In the absence of sources, Poisson's equation (10.36) reduces to Laplace's equation for the discharge potential in cylindrical coordinates

$$\frac{d}{dr}\left(r\frac{d\Phi}{dr}\right) \ = \ 0 \tag{10.37}$$

Solution of Laplace's Equation

Integrating the last expression once with respect to r, we have

$$r\, d\Phi/dr = C_1 \tag{10.38}$$

where C_1 is a constant of integration. Expression (10.38) can be rearranged to the form

$$d\Phi/dr = C_1/r \tag{10.39}$$

Integrating again, we obtain

$$\Phi(r) = C_1 \ln r + C_2 \tag{10.40}$$

where C_2 is a second constant of integration.

Expression (10.40) is a general solution to Laplace's equation (10.37) which must now be adapted to the specific case of interest. In other words, we have to determine values for the two constants of integration, C_1 and C_2 consistent with the appropriate boundary conditions.

Boundary Conditions

Earlier in this chapter, we discussed the fact that, in the vicinity of a well discharging at a constant rate Q_o, the drawdown of h is ever changing for all time, even though the flow regime in the aquifer may eventually equilibrate and the flux itself become constant. In the following, we will assume that, at a particular *snapshot* in time, equilibrium flow has been established at representative points in the aquifer, and will proceed to analyze the spatial behavior of h.

Condition 1: We assume (through conservation of flux) that whatever discharge/recharge occurs at the well head is supplied as equilibrium flow at a particular radius r. In other words, we will only select those r for which we have established equilibrium flow conditions (e.g. the difference in drawdown between two wells at different radii does not change in time). If Q_o is the total discharge/recharge at the wellhead, the aquifer recharge/discharge at r will be

$$Q_r(r) = - Q_o / 2\pi r \tag{10.41}$$

where the negative sign for Q_r is required to denote flow *toward* the origin because Q_o is implicitly positive to represent *discharge* at the wellhead.

From Darcy's law, we have that

$$Q_r = - d\Phi/dr \tag{10.42}$$

Substituting the right hand side of (10.41) for the left hand side of (10.42), cancelling negative signs and rearranging, implies that

$$d\Phi/dr = Q_o / 2\pi r \tag{10.43}$$

for any r where equilibrium flow is obtained. Applying condition (10.43) to (10.39) leads to

$$C_1 / r = Q_o / 2\pi r \tag{10.44}$$

which can be rearranged to determine

$$C_1 = Q_o / 2\pi \tag{10.45}$$

Condition 2: For a well under constant discharge Q_o starting at time $t = 0$, the radius of influence R is ever increasing during a specific well test, as long as the assumptions of our model represent the practical field situation (i.e. an ideal confined aquifer with a recharge zone at "infinity"). At any given time $t > 0$, we will assume that at and beyond some, as yet unspecified, radius of influence R, the hydraulic head is unperturbed, and assumes a value not sensibly different from the pre-existing static head of the aquifer h_o. This condition would result in the corresponding static discharge potential

$$\Phi(R) = \Phi_o = b\, K\, h_o. \tag{10.46}$$

Obviously, under constant discharge from the well Applying the condition (10.46) to (10.40) leads to

$$C_1 \ln R + C_2 = \Phi_o \tag{10.47}$$

Upon substituting in (10.47) for C_1 from (10.45), and rearranging the resultant relation, we obtain

$$C_2 = \Phi_o - Q_o \ln R / 2\pi \tag{10.48}$$

Substituting the right hand sides of (10.45) and (10.48), respectively, for C_1 and C_2 in (10.40), leads to

$$\Phi(r) = [(Q_o / 2\pi) \ln r] + \Phi_o - [(Q_o / 2\pi) \ln R] \tag{10.49}$$

which reduces to the following form for the discharge potential

$$\Phi(r) = \Phi_o + (Q_o / 2\pi) \ln (r/R) \tag{10.50}$$

or in terms of the drawdown of hydraulic potential as a function of r

$$\Phi_o - \Phi(r) = (Q_o / 2\pi) \ln (R/r) \tag{10.51}$$

Upon inspecting the right hand term in (10.51), it should be apparent that since $R \geq r$, their ratio (R/r) is always ≥ 1, and $\ln(R/r)$ is always ≥ 0. Thus, for a discharging well $(+Q_o)$, the drawdown is always positive, and $\Phi(r)$ is always $\leq \Phi_o$, the reference discharge potential at $r = R$.

Drawdown of h. We next rewrite (10.51) in terms of the hydraulic head $h(r)$ referenced to the head h_o at $r \geq R$. Since $\Phi(r) = bKh(r) + C_c$, and the aquifer transmissivity is $T = bK$, (10.51) can be expressed in the following form for the drawdown of hydraulic head

$$h_o - h(r) = [Q_o/2\pi T] \ln(R/r) \qquad (10.52)$$

Drawdown at the discharging well. We of course have a logarithmic singularity if we allow r to collapse to the origin. This can be avoided by introducing the finite radius of the well r_w. The head difference, or drawdown, at the well radius will be

$$h_o - h_w = [Q_o/2\pi T] \ln(R/r_w) \qquad (10.53)$$

where h_w is used to represent $h(r_w)$. In some cases h_w may be a useful approximation to the actual hydraulic head measured in a discharging or recharging well.

Steady-state flow in the aquifer. We point out once again for emphasis that the depth-integrated discharge in the aquifer is given by employing Darcy's law for the hydraulic potential

$$Q_r = -d\Phi/dr \qquad (10.54)$$

and operating on (10.50) to yield

$$Q_r(r) = -Q_o/2\pi r \qquad (10.55)$$

Thus the strength of the flow decreases inversely with radius — the effect of geometric spreading of flow lines is a simple consequence of the conservation of flux.

The Thiem Relation

We now come to the crux of the technique that present day hydrologists refer to as the Thiem relation. Suppose as discussed above, we have measurements of hydraulic head h in two wells at different radii r_1 and r_2, respectively. In addition, suppose that we have measures as a function of time, and determined that the flow has stabilized to steady-state condition, so that the above analysis can be applied. The hydraulic head in well one is given by

$$h_o - h_1 = [Q_o/2\pi T] \ln(R/r_1) \qquad (10.56)$$

where we have adopted the notation $h_1 = h(r_1)$. Similarly, the drawdown in well two is given by

$$h_o - h_2 = [Q_o/2\pi T] \ln(R/r_2) \qquad (10.57)$$

so that the difference, $h_2 - h_1$, is obtained by subtracting (10.57) from (10.56), resulting in

$$h_2 - h_1 = [Q_o/2\pi T][\ln(R/r_1) - \ln(R/r_2)] \qquad (10.58)$$

Recalling that the difference in the logarithms of two values is equal to the logarithm of the ratio of the values, (10.58) reduces to

$$h_2 - h_1 = [Q_o/2\pi T]\ \ln\ (r_2/r_1) \qquad (10.59)$$

This is what most modern workers refer to as the Thiem relation. If $r_2 > r_1$, then the logarithm is positive. And if Q_o is positive (i.e. the well is discharging, and flux in the aquifer is directed radially inward), we will obviously have $h_2 - h_1 > 0$.

Equation (10.59) can be rearranged to an expression for the transmissivity of the aquifer

$$T = \frac{Q_o}{2\pi\left(h_2 - h_1\right)}\ \ln\left(\frac{r_2}{r_1}\right) \qquad (10.60)$$

We apply this relation to the transient example considered above, using the difference in h shown previously in Figure 10.4.. As noted earlier, $r_1 = 50$ ft (15.24 m) and $r_2 = 100$ ft (30.48 m), and the well was discharging at a rate of 1 ft^3/min (0.472 l s^{-1}). The difference in h between these two wells at various times is used to calculate the dynamic estimate of T shown in Figure 10.7. The estimator quickly stabilizes, following the initiation of the discharge at t = 0 minutes. For example, at the end of 100 minutes (1.67 hr or 6 x 10^3 s), we obtain

$$T = 0.103\ \text{ft}^2/\text{min} = 1.59 \times 10^{-3}\ \text{m}^2\ \text{s}^{-1}$$

a value that compares favorably with the theoretical value of $T = 0.1$ ft^2/min = 1.55 x 10^{-3} m^2 s^{-1} employed in the transient model used to generate the curves in Figures 10.2, 10.3 and 10.4.

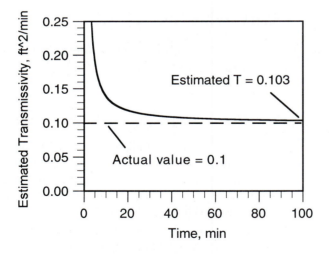

Figure 10.7 Dynamic estimation of transmissivity using Thiem relation and non-static differences in the hydraulic head measured in wells at $r_1 = 50$ ft and $r_2 = 100$ ft.

Such accuracy is not uncommon once the flow has reached steady-state. In fact, Fetter (1994) points out that the Thiem method is perhaps the most accurate way to determine the transmissivity of an aquifer; and is significantly more precise than the standard transient solutions. The disadvantage of the Thiem solution is that it does not give what is often another critical property of the aquifer: the storativity. (This is discussed in a later chapter in this text).

STEADY-STATE FLOW TO A WELL DISCHARGING FROM AN UNCONFINED AQUIFER

Statement of the Problem

Assume that a well fully penetrates an unconfined aquifer having an average hydraulic conductivity of K (Figure 10.8). The aquifer has a local saturated thickness of h, and the well pumps at a constant rate of Q_o [dimensions: $L^3 T^{-1}$]. Recall that Q_o is implicitly positive if fluid is being extracted from the aquifer and being discharged to the surface *from* the well, and is implicitly negative if fluid is being recharged *into* the well and into the aquifer. To simplify notation, we will assume for the following discussion that we are *discharging* from the well and that Q_o is implicitly positive.

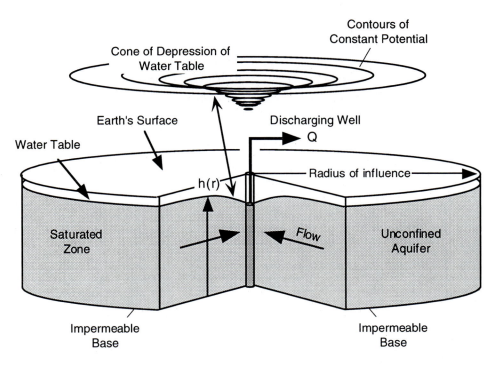

Figure 10.8 Configuration of the free surface water table in the vicinity of a well discharging from an unconfined aquifer.

Mathematical Background

We assume, for simplicity, that the base of the aquifer is horizontal, and that the Dupuit assumptions apply, so that flow is strictly horizontal. It is convenient that the elevation of the water table be referenced to the base of the aquifer. In keeping with our earlier discussion on unconfined flow, recall that the the depth-integrated discharge is given by the specific discharge q_r times the thickness of the saturated section, which for unconfined flow is the local elevation of the water table h(r), so that

$$Q_r(r) = q_r(r) \, h(r) \qquad (10.61)$$

We apply conservation of flux to the "bevelled" annulus in Figure 10.9 (bevelled because for flux in the positive radial direction h(r) will progressively decrease with r, so that the thickness of the saturated section progressively decreases as r increases).

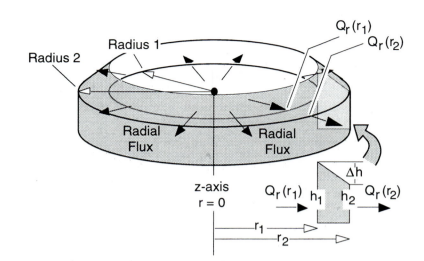

Figure 10.9 A bevelled annulus volume element used to demonstrate the consequences of flux conservation. The small inset shows Δh, the difference between the hydraulic head at two radii r_1 and r_2, respectively.

Since we assume that no fluid crosses the water table, and that the base of the aquifer is impermeable, the left hand side of our conservation integral (10.15) becomes

$$\oiint_{Surface} \mathbf{q} \cdot d\mathbf{A} = \left[q_r(r_2) \, h_2 \int_0^{2\pi} r_2 \, d\theta \right] - \left[q_r(r_1) \, h_1 \int_0^{2\pi} r_1 \, d\theta \right] \qquad (10.62)$$

which simplifies to

$$\oiint_{Surface} \mathbf{q} \cdot d\mathbf{A} = \left[q_r(r_2) \, 2\pi r_2 h_2 \right] - \left[q_r(r_1) \, 2\pi r_1 h_1 \right] \qquad (10.63)$$

Upon introducing Darcy's law for radial flow

$$q_r = - K \frac{dh}{dr} \qquad (10.64)$$

(10.63) becomes

$$\oiint_{Surface} \mathbf{q} \cdot d\mathbf{A} = - \left[K \frac{dh}{dr} \bigg|_{r_2} 2\pi r_2 h_2 \right] + \left[K \frac{dh}{dr} \bigg|_{r_1} 2\pi r_1 h_1 \right] \qquad (10.65)$$

Using Darcy's law in the form

$$Q_r(r) = - K h(r) \frac{dh}{dr} \qquad (10.66)$$

each of the square brackets on the right hand side of (10.65) can be represented by its corresponding depth-integrated discharge, or

$$\oiint_{\text{Surface}} \mathbf{q} \cdot d\mathbf{A} = \left[\, 2\pi r_2 Q_r(r_2)\,\right] - \left[\, 2\pi r_1 Q_r(r_1)\,\right] \tag{10.67}$$

This brings us full circle back to the expressions that ensued from (10.18) in our previous discussion on confined flow. That is to say, the mathematics through which we developed a differential form for the divergence condition in our earlier discussion on confined flow is strictly analogous to the present case of unconfined flow, and need not be reproduced here. It should be clear from retracing the steps leading from (10.18) through (10.31) that, in the limit of small differences between r_1 and r_2, (10.67) leads to the following differential form for the divergence of unconfined flow:

$$\frac{dQ_r}{dr} + \frac{Q_r(r)}{r} = w\, h(r) \tag{10.68}$$

where the source term on the right hand side of (10.68) is multiplied by the variable thickness of the saturated section given by $h(r)$, instead of, as in our previous discussion on confined flow, being multiplied by b, the fixed thickness of the confined aquifer.

An alternative form for (10.68) is obtained from multiplying through by r, obtaining

$$r\frac{dQ_r}{dr} + Q_r(r) = r\, w\, h(r) \tag{10.69}$$

allowing us to combine terms on the left into a single derivative such that

$$\frac{d}{dr}\left(r\, Q_r(r)\right) = r\, w\, h(r) \tag{10.70}$$

Discharge Potential for Unconfined Flow

From our definition of total discharge throughout the saturated section of the aquifer, Darcy's law can be expressed as

$$Q_r(r) = -\, Kh(r)\, \frac{dh}{dr} \tag{10.71}$$

We are prompted to define a discharge potential for unconfined flow given by

$$\Phi(r) = \frac{K\, h(r)^2}{2} + C_u \tag{10.72}$$

so that Darcy's law leads to the following form

$$Q_r = -\, \frac{\partial \Phi}{\partial r} \tag{10.73}$$

relating the total discharge to the discharge potential. (This can be seen upon substituting (10.72) into (10.73), obtaining (10.71).)

Fundamental Equations for Steady-State Unconfined Radial Flow

Substituting (10.73) for Q_r in the following divergence condition

$$\frac{d}{dr}\left(r\, Q_r(r) \right) = r\, w\, h(r) \tag{10.74}$$

we obtain the flow relation

$$\frac{d}{dr}\left(r\, \frac{\partial \Phi}{\partial r} \right) = -\, r\, w\, h(r) \tag{10.75}$$

which is a form of Poisson's equation for radial flow with local sources.

In the absence of sources, (10.75) reduces to Laplace's equation in cylindrical coordinates

$$\frac{d}{dr}\left(r\, \frac{\partial \Phi}{\partial r} \right) = 0 \tag{10.76}$$

Solution of Laplace's Equation

Integrating the latter expression (10.76) once, we have

$$r\, \partial \Phi / \partial r = C_1, \tag{10.77}$$

where C_1 is a constant of integration. Expression (10.77) can be rearranged to the form

$$\partial \Phi / \partial r = C_1 / r \tag{10.78}$$

Integrating again, we obtain

$$\Phi(r) = C_1 \ln r + C_2 \tag{10.79}$$

where $\ln r$ is the natural logarithm of r to the base e, and C_2 is a second constant of integration.

Expression (10.79) is a general solution to Laplace's equation (10.76) which must now be adapted to the specific case of interest. In other words, we have to determine values for the two constants of integration, C_1 and C_2, that are consistent with our present problem of interest. This is done by applying the appropriate boundary conditions.

Boundary Conditions

As in the case of confined flow described earlier in this chapter, the hydraulic head h in the vicinity of a well discharging (or recharging) at constant Q_o will be ever decreasing (or increasing) with time. However, after a preliminary transient phase, it is quite reasonable to analyze equilibrium or steady-state flow at representative points in the aquifer at specific snapshots in time. In practice this equilibrium flow will be identified by monitoring the difference in hydraulic head $\Delta h = h_2 - h_1$, between two wells, respectively, as a function of time. When Δh stabilizes, we assume the hydraulic gradient, hence the depth-integrated discharge Q_r, is stabilized within the radius monitored.

Boundary condition 1: We will assume that whatever discharge/recharge Q_o occurs at the well head at $r = 0$ is supplied as equilibrium flow at a particular radius r. The depth integrated flux at r will be

$$Q_r(r) = - Q_o / 2\pi r \qquad (10.80)$$

where the negative sign for Q_r is required to denote flow toward the origin to account for *discharge* at the wellhead. Substituting from Darcy's law

$$Q_r = - d\Phi/dr \qquad (10.81)$$

for Q_r on the left side of (10.81), we have

$$d\Phi/dr = Q_o / 2\pi r \qquad (10.82)$$

for any r where equilibrium flow is obtained. Thus differentiating (10.79) with respect to r, and setting the resulting expression equal to the right side of (10.82), then cancelling common terms on either side of the equation leads to

$$C_1 = Q_o / 2\pi \qquad (10.83)$$

so that (10.79) becomes

$$\Phi(r) = [Q_o / 2\pi] \ln r + C_2 \qquad (10.84)$$

Boundary condition 2: We will assume, for a well under constant discharge Q_o starting at time $t = 0$, that at and beyond some radius of influence R (depending on the time), the static discharge potential is unperturbed from its pre-pumping static value, which using the definition of the discharge potential from (10.72), and setting the arbitrary constant $C_u = 0$, implies

$$\Phi(R) = \Phi_o = K h_o^2 / 2 \qquad (10.85)$$

(We recognize, of course, that under constant discharge from the well, the radius of influence R is ever increasing during a specific well test.)

Applying the left equality in (10.85) to (10.84), and rearranging the resultant relation, we obtain

$$C_2 = \Phi_o - Q_o \ln R / 2\pi \qquad (10.86)$$

Substituting C_2 from (10.86) into (10.84), after rearranging, leads to the following form for the discharge potential

$$\Phi(r) = \Phi_o + [Q_o / 2\pi] \ln (r/R) \qquad (10.87)$$

which can be written in terms of the drawdown of hydraulic potential as a function of r

$$\Phi_o - \Phi(r) = [Q_o / 2\pi] \ln (R/r) \qquad (10.88)$$

Drawdown of h. By recalling the definition of the discharge potential for unconfined flow given by

$$\Phi(r) = \frac{K\,h(r)^2}{2} + C_u \qquad (10.89)$$

where for the present case we can set $C_u = 0$, we can rewrite (10.88) in terms of the hydraulic head $h(r)$, or

$$h_o{}^2 - h(r)^2 = \left[Q_o / \pi K \right] \ln(R/r) \qquad (10.90)$$

Upon rearranging and taking the square root, we have

$$h(r) = \sqrt{ h_o{}^2 - \left[Q_o / \pi K \right] \ln(R/r) } \qquad (10.91)$$

Drawdown at the discharging well. As in the case of the confined aquifer discussed above, in order to calculate the drawdown at the well while avoiding the logarithmic singularity as r approaches the origin, we introduce the finite radius of the well r_w. The head difference (or drawdown) at the well radius will thus be

$$(h_o)^2 - (h_w)^2 = [Q_o / \pi K] \ln(R/r_w) \qquad (10.92)$$

where h_w is used to represent $h(r_w)$.

Defining an "average" transmissivity. We can decompose (10.92) into the following form

$$(h_o - h_w)(h_o + h_w) = [Q_o/\pi K] \ln(R/r_w). \qquad (10.93)$$

Dividing both sides of the expression by 2, and defining the "average" saturated thickness of the aquifer according to

$$h_{avg} = (h_o + h_w)/2 \qquad (10.94)$$

we can define an *average transmissivity* as

$$T_{avg} = K\,h_{avg} \qquad (10.95)$$

Expression (10.93) can thus be rewritten in the form

$$(h_o - h_w) = [Q_o/(2\pi T_{avg})] \ln(R/r_w) \qquad (10.96)$$

which is identical to the expression for the drawdown of a confined aquifer discussed in an earlier section providing T for the confined case is replaced by T_{avg} as defined by (10.95) above.

A form of Darcy's law. Again decomposing (10.92) to the expression (10.93), we can rearrange it to the form

$$Q_o = 2\pi \left[\frac{h_o + h_w}{2} \right] \left[K\,\frac{h_o - h_w}{\ln(R/r_w)} \right] \qquad (10.97)$$

- 138 -

so that the total well discharge can be viewed as a modified version of Darcy's law. To see this more clearly, allow R to approach r_w. Consider the annulus between $r_w = R - \Delta r$ and R. The ratio R/r_w is thus approximately given by

$$R / r_w \cong (1 + \Delta r/R) \tag{10.98}$$

Thus, using a Taylor's series expansion

$$\ln (R / r_w) \cong \Delta r/R - (\Delta r/R)^2 / 2 + (\Delta r/R)^3 / 3 - (\Delta r/R)^4 / 4 + ... \tag{10.99}$$

So that, we can write

$$Q_o = 2 \pi R \left[\frac{h_o + h_w}{2} \right] \left[K \frac{h_o - h_w}{\Delta r} \right] \tag{10.100}$$

which is, perhaps, a more obvious form of Darcy's law, where the cross sectional area of flow

$$\text{AREA} = 2 \pi R \left[\frac{h_o + h_w}{2} \right] \tag{10.101}$$

is given as the circumference $2\pi R$ times the average thickness of the saturated section given by $(h_o + h_w)/2$.

If we define the length of the flow path as

$$L = \Delta r \tag{10.102}$$

then we can rewrite (10.100) as

$$Q_o = \text{AREA} \times K \frac{h_o - h_w}{L} \tag{10.103}$$

where the lack of the usual negative sign on the right hand side is in keeping with the fact that the flow of a well is usually considered positive when it is discharging, or when flow is in the negative radial direction.

Steady-state flow in the aquifer. The depth-integrated discharge for unconfined flow in the aquifer is given, from the conservation of flux, by equating (10.61) to the total discharge from (10.80), such that

$$Q_r (r) = q_r (r) h(r) = - Q_o / 2\pi r \tag{10.104}$$

which can be rearranged to give us an expression for the specific discharge

$$q_r(r) = - \frac{Q_o}{2 \pi r h(r)} \tag{10.105}$$

As in the case of confined flow, the strength of the flow decreases inversely with radius as a simple consequence of geometric spreading of flow lines; but, in addition, flow decreases inversely with the thickness of the saturated section $h(r)$, which upon substituting (10.91) for $h(r)$ in (10.105), we have the following form for specific discharge

$$q_r(r) = - \frac{Q_o}{2 \pi r \sqrt{h_o^2 - \left[Q_o / \pi K \right] \ln (R / r)}} \tag{10.106}$$

Alternatively, the latter expression may be obtained directly by beginning with Darcy's law

$$q_r(r) = - K \frac{dh(r)}{dr} \tag{10.107}$$

and differentiating (10.91) with respect to r.

The Thiem Relation for Unconfined Flow

Steady-state flow relations. The Thiem relation for unconfined flow can be derived in a parallel fashion to the relation for confined flow described above. Again, we suppose we have measurements of hydraulic head h in two wells at different radii r_1 and r_2, respectively. In addition, we assume that the flow — not necessarily the hydraulic head — has stabilized to steady-state condition. The hydraulic head in well #1 at radius r_1 is given by

$$h_o^2 - h_1^2 = \left[Q_o / \pi K \right] \ln (R / r_1) \tag{10.107}$$

where, as before, we have adopted the abbreviated notation $h_1 = h(r_1)$. Similarly, the drawdown in well #2 at radius r_2 is given by

$$h_o^2 - h_2^2 = \left[Q_o / \pi K \right] \ln (R / r_2) \tag{10.108}$$

so that the difference is obtained by subtracting (10.108) from (10.107), resulting in

$$h_2^2 - h_1^2 = \left[Q_o / \pi K \right] \ln (R / r_1) - \left[Q_o / \pi K \right] \ln (R / r_2) \tag{10.109}$$

Employing the fact that the difference in the logarithms of two values is equal to the logarithm of the ratio of the values, (10.109) reduces to

$$h_2^2 - h_1^2 = \left[Q_o / \pi K \right] \ln (r_2 / r_1) \tag{10.110}$$

This is the Thiem relation for equilibrium unconfined flow. If $r_2 > r_1$, then the logarithm is positive. And if Q_o is positive (i.e. the well is discharging, and flux in the aquifer is directed radially inward), we will have $h_2^2 - h_1^2 > 0$, so that a "cone of depression" forms around the well. Expression (10.110) can be rearranged to an expression for the hydraulic conductivity of the aquifer

$$K = \frac{Q_o}{\pi \left(h_2^2 - h_1^2 \right)} \ln \left(\frac{r_2}{r_1} \right) \tag{10.111}$$

Note that Thiem's relation for unconfined flow leads to an expression for the conductivity, whereas in confined flow it leads to an expression for the transmissivity (see (10.60)). Again, we can draw a parallel with our previous discussion on confined flow. We define an average thickness for the saturated section between the two monitoring wells by

$$b_{avg} = h_{avg} = \frac{(h_2 + h_1)}{2} \tag{10.112}$$

where we use the notation b_{avg} as an alternative to h_{avg} to make the analogy with the confined flow case more apparent.

If we next define an average transmissivity for unconfined flow by

$$T_{avg} = b_{avg} K \tag{10.113}$$

then multiplying both sides of (10.111) by (10.112), the resulting expression becomes

$$T_{avg} = \frac{Q_o}{2\pi(h_2 - h_1)} \ln\left(\frac{r_2}{r_1}\right) \tag{10.114}$$

which can be interpreted as the "average transmissivity" for unconfined flow in the saturated section between the two wells at radii r_1 and r_2, respectively; and is the analog of expression (10.60) for confined flow.

Note on a common unconfined flow approximation. In the expression that we have developed for unconfined flow to a discharging well

$$h(r) = \sqrt{h_o^2 - [Q_o/\pi K] \ln(R/r)} \tag{10.91}$$

the transmissivity of the saturated section

$$T = K h(r) \tag{10.115}$$

is implicitly assumed to be a function of radius, since the thickness of the saturated section h is, itself, a function of radius. Some authors, however, either for the steady-state case or the transient case to be considered in a later chapter, assume that the transmissivity of the saturated section is simply

$$T_o = K h_o \tag{10.116}$$

where h_o is the static level of the water table (or the elevation of the unperturbed head at some distance from the discharging/recharging well). In a sense, this is equivalent to the confined flow assumption where it is assumed that flow is uniformly distributed over a fixed thickness b, where here b is replaced by the thickness of the undisturbed, pre-pumping saturated zone. These authors do this for the purpose of invoking an approximation to unconfined flow of the form

$$h(r) = h_o - [Q_o/2\pi T_o] \ln(R/r) \tag{10.117}$$

which is a simple modification of the rearranged confined flow result given by (10.52), with T replaced by the approximation T_o given by (10.116). While this is more commonly done for transient flow than for the steady-state case (because of a significant computational advantage), it is instructive to consider the implications of the approximation here for equilibrium flow.

We begin by factoring the more general expression (10.91) to the form

$$h(r) = h_o \sqrt{1 - \left[Q_o / \left(h_o^2 \pi K \right) \right] \ln (R/r)} \qquad (10.118)$$

Combining one of the h_o factors in the denominator under the radical sign with K according to (10.116) above, we obtain

$$h(r) = h_o \sqrt{1 - \left[Q_o / \left(h_o \pi T_o \right) \right] \ln (R/r)} \qquad (10.119)$$

We next investigate an asymptotic form of (10.119) in the limit of

$$\left[Q_o / \left(h_o \pi T_o \right) \right] \ln (R/r) \ll 1.0 \qquad (10.120)$$

We can use the binomial theorem

$$\sqrt{1 - \varepsilon} \approx 1 - \frac{\varepsilon}{2} \qquad (10.121)$$

to reduce expression (10.119) to the form

$$h(r) = h_o - \left[Q_o / 2 \pi T_o \right] \ln (R/r) \qquad (10.122)$$

which is precisely the relation obtained above where we simply replaced T = b K for a *confined* aquifer by the approximation T = K h_o for an *unconfined* aquifer. To get from (10.91) to (10.122) analytically, we had to assume (10.120). This has some significant undertones for practical applications. Recalling our expression for drawdown of an unconfined aquifer

$$h_o^2 - h(r)^2 = \left[Q_o / \pi K \right] \ln (R/r) \qquad (10.90)$$

we divide both sides by h_o^2, and upon substituting $T_o = K h_o$ on the right hand side, we obtain

$$\frac{h_o^2 - h(r)^2}{h_o^2} = \left[\frac{Q_o}{h_o \pi T_o} \right] \ln \left(\frac{R}{r} \right) \qquad (10.123)$$

Clearly the right hand side of (10.123) is equivalent to the left hand side of the inequality (10.120), leading to

$$\frac{h_o^2 - h(r)^2}{h_o^2} \ll 1.0 \qquad (10.124)$$

which, in turn is approximated by the condition

$$2 \frac{h_o - h(r)}{h_o} \ll 1 \qquad (10.125)$$

(The factor of 2 in the numerator of the latter expression comes from a binomial approximation to the squared terms in (10.124)).

Note: Conditions (10.124) or (10.125) are quite stringent requirements that are commonly violated by practical well tests, and in the vicinity of discharging production wells.

PART III. TRANSIENT FLOW

Chapter 11. Introduction to Transient Flow

FUNDAMENTAL RELATIONS

Causes of Transient Flow

A transient response in the temporal and spatial character of the flow field will be associated with any change in the hydraulic parameters of an aquifer, such as

- An imposed change in hydraulic head,
- A modification in fluid influx in the recharge or discharge regions,
- A change in the storage conditions,
- An increase or decrease in the discharge from wells, etc.

In this section we discuss some of the underlying physical processes that govern such transient behavior.

Conservation of Fluid Flux: Continuity Condition with Sources

The net fluid flux through the surface of a unit volume (the rate at which a fluid transports material *out* of a unit volume minus the rate at which a fluid transports material *into* a finite volume) must equal the rate at which material is being created or is *decreasing* in the volume.

$$\oiint_{\text{surface}} \mathbf{q} \cdot d\mathbf{A} = \iiint_{\text{volume}} \left(r - \frac{dS_w}{dt} + w_g \right) dV \qquad (11.1)$$

where

$\mathbf{q}$ is the material flux transported by the fluid out of the volume,
r is the local recharge or infiltration from precipitation, streams or other sources,
dS_w/dt is the change in the volume of water stored per unit volume of the host material. The negative sign is called for since a *decrease* in the material stored within a volume is associated with a positive material flux *out* of the volume.
w_g is a generalized water source to account for other contributions, not specifically named here.

Transient Effects from the Aquifer

The source terms r and w_g in (11.1) represent effects external to the aquifer. Here we want to explore source contributions from the aquifer itself, and these are largely in the guise of induced changes in the storage of material — the term dS_w/dt in (11.1). In this case the expression reduces to the form

$$\oiint_{\text{surface}} \mathbf{q} \cdot d\mathbf{A} = -\iiint_{\text{volume}} \frac{dS_w}{dt} dV \qquad (11.2)$$

where, as above, S_w is the volume of water stored per unit volume of the host material that is able to change with time.

In the case of a small stationary volume, following the discussion in Chapter 5 leading up to (5.33) and (5.34), expression (11.2) leads to the divergence condition:

$$\nabla \cdot \mathbf{q} = -\partial S_w/\partial t \tag{11.3}$$

or

$$\partial q_x/\partial x + \partial q_y/\partial y + \partial q_z/\partial z = -\partial S_w/\partial t \tag{11.4}$$

which in one dimension becomes simply

$$\partial q_x/\partial x = -\partial S_w/\partial t \tag{11.5}$$

In addition to (11.3), we have Darcy's law,

$$\mathbf{q} = - K\nabla h, \tag{11.6}$$

where we have assumed that the medium is isotropic.

In one dimension, Darcy's law (11.6) becomes

$$q_x = - K \, \partial h/\partial x \tag{11.7}$$

Upon substituting the right hand side of (11.7) for q_x in the divergence condition (11.5), and assuming the medium is homogeneous (i.e. K is not a function of position), we obtain the following expression for transient saturated flow in one dimension:

$$\partial^2 h/\partial x^2 = (1/K)\partial S_w/\partial t \tag{11.8}$$

In three dimensions, upon substituting (11.6) into (11.3), our flow relation becomes

$$\partial^2 h/\partial x^2 + \partial^2 h/\partial y^2 + \partial^2 h/\partial z^2 = (1/K)\partial S_w/\partial t \tag{11.9}$$

Our attention is directed now to the case where temporal changes in the stored water S_w are induced by transient changes in subsurface conditions.

Transient Response to a Discharge Event

A fundamental process of subsurface flow is the release and/or storage of water in aquifers due to transient changes in ambient conditions. The physical processes involved in a discharge event are

- Discharge from the saturated section of an aquifer,
- Dewatering or gravity drainage of the saturated section,
- Elastic deformation of the saturated section,
- Physical decompression of water,
- Dewatering of chemically "bound" water — usually bound to clays,
- "Recharge" of the saturated section from distant sources.

Compressibility

Consider the volume V in Figure 11.1 subject to a pressure increase Δp.

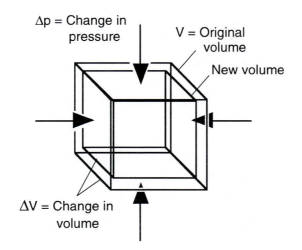

Figure 11.1 Change in volume (ΔV decreasing) due to change in pressure (Δp increasing).

The volume will change by an amount ΔV given by

$$\Delta V = - \text{Constant} \cdot V \cdot \Delta p \tag{11.10}$$

In differential form, this becomes

$$dV/dp = - \text{Constant} \cdot V \tag{11.11}$$

Compressibility of water. We use the notation of β for the compressibility of water given by

$$\beta = - (1/V_w)dV_w/dp \tag{11.12}$$

where V_w is the volume of the sample of water being compressed, dV_w is the incremental change in volume, and dp is the increment in pressure. Clearly β will have the units of inverse pressure $[m^2 \ N^{-1}]$.

An alternative form to (11.12) is

$$\beta = (1/\rho_w)d\rho_w/dp \tag{11.13}$$

where ρ_w is the density of the fluid. (Note the negative sign for (11.12) and the implicit positive sign for (11.13), in keeping with the fact that a *decrease* in volume is consistent with an *increase* in density.

Compressibility of the host matrix. We employ the notation α for the compressibility of the matrix or host material containing the water. Assuming for now that the interstitial water offers no counter-stress (i.e. fluid freely flows out of or into the matrix volume under stress), in general

$$\alpha = - (1/V_t)dV_t/d\sigma \qquad (11.14)$$

where

V_t is the total volume of the matrix sample

σ is the applied stress or force per unit area

The matrix compressibility α will have the units of inverse stress, or the inverse of force per unit area [$m^2 \, N^{-1}$].

Compaction. Assume that an external stress (σ_{ext}) is applied to a unit volume of saturated material. Compaction might occur due to one or all of three mechanisms

1. By compression of the aquifer matrix.
2. By compression of the interstitial water.
3. By rearrangement of the matrix material (rearrangement of sand grains, movement along fractures, etc.).

COMPRESSIBILITY OF A FLUID SATURATED POROUS MEDIUM

Experimentally Determining α

Consider, as shown in Figure 11.2, a piston cylinder apparatus containing a water-saturated sample of soil (although any porous sample would do), with the pistons configured to allow water to freely escape into a holding reservoir. We assume that, at the scale of this experiment, we can neglect gravitational effects.

It should be intuitive that when fluid is able to freely enter and leave such a porous sample, the compressibility of the composite bulk material is controlled solely by the properties of the matrix, because in this case the fluid is free to flow unrestrained, and offers zero *back pressure*.

Assume that b is the original length of the sample, and that its original porosity is n, where

$$n = V_v/V_t \qquad (11.15)$$

where

V_t is the total volume of the sample,
V_v is the volume of void space.

The compressibility can be determined at each applied stress σ (force/unit area), by using

$$\alpha = - (1/b)[db/d\sigma] \qquad (11.16)$$

or

$$\alpha = - (1/V_t)[dV_t/d\sigma] \qquad (11.17)$$

where

$$V_t = V_s + V_v \qquad (11.18)$$

[The total volume V_t is the sum of the volume of the material in the solid matrix V_s and the interstitial voids V_v.]

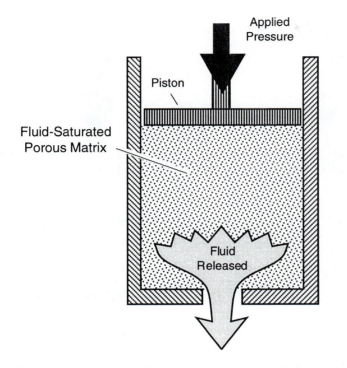

Figure 11.2 Piston pressure apparatus with an outlet for pore fluid.

While, in general

$$dV_t = dV_s + dV_v \qquad (11.19)$$

usually the compressibility of the individual soil grains or the rock fabric, itself, is inconsequential, so dV_s is negligible. Thus, we assume that any deformation is associated with a change in the void volume, or

$$dV_t \approx dV_v \qquad (11.20)$$

Note: In practice, α as determined from (11.17), and using an apparatus of the type illustrated in Figure 11.2, is neither constant nor is the measurement reversible. Permanent, irreversible deformations are the rule. Usually α is not a constant coefficient as implied by (11.17), but is a function of the applied stress, as well as the prior deformation history of the sample. Moreover, α is usually dependent on the *sign* of the change in stress — it is not atypical for the compressibility to be significantly larger during compression (loading) than during expansion (unloading). For clays, this might be a factor of 10:1; but for uniform sands, it might be closer to 1:1. Thus, (11.17) should be primarily thought of as a qualitative conceptual paradigm, and the strict linearity that it implies between stress and strain at best pertains to a limited range of values that, in turn, depend on the material being investigated.

Effective Stress

Consider the experimental arrangement shown in Figure 11.3. The vessel on the left (pressure vessel #1) is exactly the same as the piston pressure apparatus in Figure 11.2, except that vessel #2 can offer a "back pressure" to prevent the fluid from freely flowing from #1. Assume pressure vessel #1 is filled with water saturated sand, and vessel #2 is simply filled with water, serving as a pressurized reservoir. Piston #1 bears down with stress σ_1 and piston #2 bears down with stress σ_2 (again the stress here is normal force per unit area).

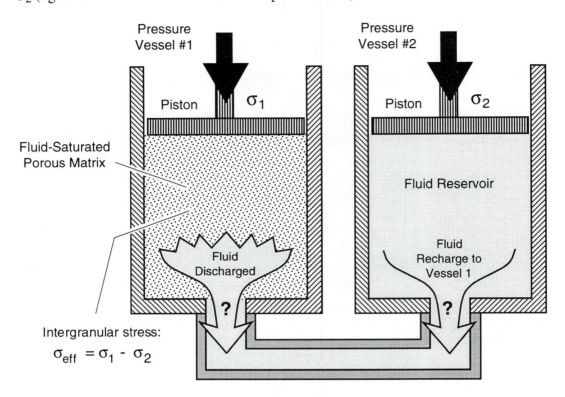

Figure 11.3 Pressure vessel #1 on the left, containing a fluid saturated sample, is hydraulically connected to pressure vessel #2 on the right that can offer a "back pressure" to vessel #1. This back pressure results in a pore pressure in the sample in vessel #1, that in turn modifies the effective stress on the specimen.

Assuming, for now, that both σ_1 and σ_2 are *externally* applied (and neglecting internal gravitational body forces of the form ρg), the *effective stress* acting on medium #1 is given by

$$\sigma_{eff} = \sigma_1 - \sigma_2 \qquad (11.21)$$

If $\sigma_1 = \sigma_2$ then the two stresses are just balanced out, and there is no deformation of medium #1. If $\sigma_2 = 0$, then the total stress of σ_1 is imparted to deform medium #1.

Now consider a more geologically relevant situation, where the stresses act across a plane at some depth d. Define the *geologic stress* σ_{geol} representing the sum total of all geologic and gravitational stresses acting on the column of rock and water above the interface. Note that, in this case, we *do not* neglect gravitational body forces of the form ρg. (Some authors refer to this "geologic" stress synonymously as the "lithologic" or "tectonic" stress.) In addition to the geologic stress transmitted through the contact between soil grains or the rock fabric, assume that we have a pressure p associated with an interstitial fluid — the so-called "pore pressure". If the pores of the material at depth d are open and connected to atmospheric pressure at the surface, the fluid will be at hydrostatic pressure

$$p = \rho\, g\, d \qquad (11.22)$$

Or, the fluid might be sealed off from the surface and be "overpressured", or "underpressured" relative to its expected hydrostatic value. In any case, the pressure will be associated with some "pressure head" h_p

$$p = \rho\, g\, h_p \qquad (11.23)$$

as discussed earlier earlier in the text, but where, in general,

$$h_p \neq d \qquad (11.24)$$

The "effective stress" is the actual stress responsible for compaction of the interstices or voids, and is the difference between σ_{geol} and the pressure p acting on the interstitial pore fluid. This differential pressure (termed the effective stress or the effective pressure) is defined by

$$\sigma_{eff} = \sigma_{geol} - p \qquad (11.25)$$

Or, for infinitesimal changes

$$d\sigma_{eff} = d\sigma_{geol} - dp \qquad (11.26)$$

It often happens that the geologic stress σ_{geol} does not vary appreciably throughout the characteristic time interval of interest: a few hours for a well test, a few weeks for the infiltration of a storm event, or, on a regional scale, even many years for a climatic variation. Thus, it is not unreasonable to assume that, for a range of useful applications,

$$d\sigma_{geol} = 0 \qquad (11.27)$$

so that

$$d\sigma_{eff} = - dp \qquad (11.28)$$

The latter expression implies that, as the fluid pressure *increases*, the effective stress *decreases* by an equivalent amount.

Effective Stress in Terms of Hydraulic Head

Recall that the pressure at a point in a saturated medium is given by

$$p = \rho\, g\, h_p = \rho\, g\, \psi \qquad (11.29)$$

where $h_p = \psi$ is the pressure head, given by

$$\psi = h - z \qquad (11.30)$$

h being the hydraulic head and z being constant for a specific reference point.

Substituting for ψ, we have

$$p = \rho\, g\, (h - z)$$ (11.31)

By differentiating (11.31), and substituting dp into the right side of (11.28), we have that

$$d\sigma_{eff} = -\rho\, g\, dh$$ (11.32)

which illustrates that changes in the effective pressure are simply related to changes in the hydraulic head.

CONFINED AQUIFERS: MECHANISMS OF WATER RELEASE

Importance of the Elastic Properties of an Aquifer

Recall the form of the flow relation from an earlier discussion in the chapter, repeated here for convenience:

$$\partial^2 h/\partial x^2 + \partial^2 h/\partial y^2 + \partial^2 h/\partial z^2 = (1/K)\partial S_w/\partial t$$ (11.9)

In this section, we want to apply (11.9) to the case of a confined aquifer, and to relate the transient storage term on the right hand side to the elastic properties — the compressibility — of the aquifer matrix and the aquifer fluid itself. We will see that the finite compressibility of such a material is fundamentally responsible for the character of its transient response to a change in its hydraulic condition. If both the host matrix and the interstitial fluid were incompressible, then, upon pumping, for example, a new equilibrium condition would be established instantaneously throughout the aquifer (Strack, 1989) — and there would be no transient effect.

As a preamble to the following discussion, we note that for a confined aquifer, a decrease in the hydraulic head leads to the release of water from storage through two mechanisms:

1. According to (11.32) a *decrease* in h is associated with an *increase* in the effective stress. The increase in effective stress, according to (11.17) causes a *decrease* in the interstitial volume and the resulting compaction of the aquifer matrix. In a sense, water is lost from storage as it is "squeezed out" of the interstitial volume as the pore pressure loses ground to the compressive stress on the matrix.

2. A decrease in h is related to a decrease in the fluid pressure, which according to (11.12) will cause an expansion of the water, so that the total water volume effectively *increases*, appearing as an incremental release of fluid from the original fixed reference volume of the host matrix.

We consider the relation of these two mechanisms to the transient storage term on the right hand side of (11.9) in more detail below.

Release of Water from Storage by Compaction

Relation (11.17) can be rewritten to give the change in volume dV_t of a material due to a change in the effective stress $d\sigma_{eff}$ in the form

$$dV_t = -\alpha V_t\, d\sigma_{eff}$$ (11.33)

where α is the matrix compressibility. Recall also that the compressibility of the individual soil grains or the rock fabric, itself, is inconsequential, so that we can assume that any deformation in the host matrix is associated with a change in the void volume, or, repeating from above

$$dV_t \approx dV_v \tag{11.20}$$

Clearly, if the effective stress increases, the void volume decreases, with the consequence that any water within that void volume (we assume 100% saturation) will be expelled — in other words, lost from storage. If we define the term $dV_w^{(\alpha)}$ as the increment of water released from the reference volume due to compression of the host matrix, we would have that

$$dV_w^{(\alpha)} = -dV_v \tag{11.34}$$

which upon substituting into (11.20), and then into (11.33), results in the following expression for the water expelled from a total volume V_t due to compaction of the host matrix:

$$dV_w^{(\alpha)} = \alpha V_t \, d\sigma_{eff} \tag{11.35}$$

This relation states that a positive $d\sigma_{eff}$ causes a compression of the bulk matrix, with a concomitant expulsion of water ($+dV_w^{(\alpha)}$) from the matrix.

We next recall the relation from above that relates a change in hydraulic head to a change in effective stress, repeated here for convenience as

$$d\sigma_{eff} = -\rho \, g \, dh \tag{11.32}$$

Substituting the right hand side of the latter expression for $d\sigma_{eff}$ in (11.35), we obtain

$$dV_w^{(\alpha)} = -\alpha V_t \, \rho \, g \, dh \tag{11.36}$$

which relates the expulsion of water from an elastic matrix due to a change in hydraulic head. The mechanism is that a decrease in hydraulic head is associated with an increase in the effective stress. The latter causes a decrease in the void volume which, in turn, is responsible for the expulsion of water from the reference volume. By convention this represents a net production of water (a positive supply) from the reference volume.

Release of Fluid Volume through Decompression

Rearranging (11.12) leads to the following expression for the volumetric change of water due to its compression:

$$dV_w^{(\beta)} = -\beta V_w \, dp \tag{11.37}$$

where

$dV_w^{(\beta)}$ is the increment of water released from the reference volume due to its own decompression,

β is the fluid compressibility,

V_w is the volume of water present in the total volume of the reference element,
dp is the increment of pressure applied to the reference volume.

A positive dp causes a volumetric compression of the fluid $(- dV_w^{(\beta)})$. Recall from (11.28) above, however, that an incremental *increase* in fluid pressure is related to an incremental *decrease* in the effective stress, or $dp = - d\sigma_{eff}$, thus (11.37) can be rewritten in the form

$$dV_w^{(\beta)} = \beta V_w d\sigma_{eff} \tag{11.38}$$

In words, (11.38) states that a positive $d\sigma_{eff}$ is related to a decompression of the fluid in the interstitial pore spaces in direct proportion to the volume of water present V_w and its compressibility. Since this excess water volume $dV_w^{(\beta)}$ has to move someplace, there is a concomitant expulsion of water $(+dV_w^{(\beta)})$ from the matrix. If, in fact, water were not allowed to flow from the matrix, it would exert a corresponding back pressure to mitigate the change in $d\sigma_{eff}$.

Expression (11.38) gives the incremental water released dV_w from a total water volume of V_w. Of course, water does not comprise 100% of our reference volume; the material making up the host matrix contributes to this volume also. Recall from earlier in the chapter that the porosity n is given by

$$n = V_v / V_t \tag{11.15}$$

where

V_t is the total volume of the sample,
V_v is the volume of void space.

In a saturated sample, water will fill the void space, so that $V_w = V_v$, which upon substituting into (11.15), and rearranging, leads to $V_w = n V_t$. Substituting the latter into (11.38), leads to

$$dV_w^{(\beta)} = n \beta V_t d\sigma_{eff} \tag{11.39}$$

Finally, substituting the right hand side of (11.32) for $d\sigma_{eff}$ in (11.39), we have

$$dV_w^{(\beta)} = - n \beta V_t \rho g dh \tag{11.40}$$

which stated in words states that a decrease in hydraulic head is responsible to a decrease in pore pressure, in turn leading to a decompression of the interstitial fluid. This volume fraction expands, and is responsible for the local production of water in the reference volume.

Total Water Released as Specific Storage

The total incremental volume of water released is given by the sum

$$dV_w = dV_w^{(\alpha)} + dV_w^{(\beta)} \tag{11.41}$$

Substituting (11.36) and (11.40) for the respective terms in (11.41), we obtain

$$dV_w = - [\alpha V_t + n\beta V_t] \rho g dh, \tag{11.42}$$

where ρ is the density of water, g is the local force of gravity and dh is an incremental increase in hydraulic head.

From (11.42) we see that a decrease in head is associated with an expulsion of water ($+dV_w$) from the matrix interstices as well as an increase in its volume through decompression; both effects represent a decrease in storage, and contribute positively to the production of the volume of free water.

Definition: The *specific storage* of an aquifer is the relative volume of stored water released through elastic deformation from (or absorbed by) a volume of an aquifer through a decrease (or increase) of the hydraulic head.

Normalizing both sides of (11.42) by the total volume, V_t, and the decrease in hydraulic head, dh, we obtain the following expression for the *specific storage*

$$S_s = [\alpha + n\beta]\,\rho g \qquad (11.43)$$

Since α and β have units of $m^2\,N^{-1}$, n is dimensionless, and ρg, the body force, has units of $N\,m^{-3}$, the units of the specific storage S_s are simply m^{-1}, which is consistent with the above definition.

Therefore, the volume fraction of water lost from a volume of the aquifer is related to the specific storage by

$$dV_w/V_t = -S_s\,dh \qquad (11.44)$$

Assuming that the incremental changes, dV_w and dh, occur in a time increment dt, we have, at a fixed point in the aquifer, in the limit of small dt

$$\frac{1}{V_t}\frac{\partial V_w}{\partial t} = -S_s\frac{\partial h}{\partial t} \qquad (11.45)$$

which is the fractional volume of water expelled from an elemental volume of the aquifer per unit time.

Recall that the fractional volume of water expelled from a volume of the aquifer per unit time is the negative of the time rate of change of the water stored in the volume, or

$$\frac{1}{V_t}\frac{\partial V_w}{\partial t} = -\frac{\partial S_w}{\partial t} \qquad (11.46)$$

Thus, upon equating the right hand sides of (11.45) and (11.46), we obtain

$$\frac{\partial S_w}{\partial t} = S_s\frac{\partial h}{\partial t} \qquad (11.47)$$

which relates the time rate of change in storage to the specific storage and the time rate of change of hydraulic head.

CONFINED AQUIFERS: THE TRANSIENT FLOW EQUATION

Diffusion Equation for Hydraulic Head, Specific Storage and Hydraulic Conductivity

Assuming an isotropic medium, one may substitute the right hand side of Darcy's law in three dimensions given by (11.6) for the specific discharge **q** on the left hand side of the divergence relation (11.3), the following general form for the transient flow equation applies to both confined or unconfined aquifers:

$$\left[\frac{\partial}{\partial x}\left(K\frac{\partial h}{\partial x} \right) + \frac{\partial}{\partial y}\left(K\frac{\partial h}{\partial y} \right) + \frac{\partial}{\partial z}\left(K\frac{\partial h}{\partial z} \right) \right] = \frac{\partial S_w}{\partial t} \tag{11.48}$$

For a confined aquifer, where the only source or sink of water is the release (or take up) of fluid by local elastic deformation, substituting (11.47) for the transient source term on the right hand side of (11.48) leads to

$$\left[\frac{\partial}{\partial x}\left(K\frac{\partial h}{\partial x} \right) + \frac{\partial}{\partial y}\left(K\frac{\partial h}{\partial y} \right) + \frac{\partial}{\partial z}\left(K\frac{\partial h}{\partial z} \right) \right] = S_s \frac{\partial h}{\partial t} \tag{11.49}$$

This is the fundamental transient flow relation for the hydraulic head in a heterogeneous, isotropic confined aquifer. For strictly horizontal flow, (11.49) reduces to the two dimensional form

$$\frac{\partial}{\partial x}\left(K\frac{\partial h}{\partial x} \right) + \frac{\partial}{\partial y}\left(K\frac{\partial h}{\partial y} \right) = S_s \frac{\partial h}{\partial t} \tag{11.50}$$

or the one dimensional form

$$\frac{\partial}{\partial x}\left(K\frac{\partial h}{\partial x} \right) = S_s \frac{\partial h}{\partial t} \tag{11.51}$$

Assuming the medium is homogeneous and isotropic, the diffusion equation for transient flow in 3-D in a confined aquifer becomes

$$\frac{\partial^2 h}{\partial x^2} + \frac{\partial^2 h}{\partial y^2} + \frac{\partial^2 h}{\partial z^2} = \frac{S_s}{K}\frac{\partial h}{\partial t} \tag{11.52}$$

which, for horizontal flow reduces to the 2-D form

$$\frac{\partial^2 h}{\partial x^2} + \frac{\partial^2 h}{\partial y^2} = \frac{S_s}{K}\frac{\partial h}{\partial t} \tag{11.53}$$

For one dimensional flow, (11.52) reduces to

$$\frac{\partial^2 h}{\partial x^2} = \frac{S_s}{K}\frac{\partial h}{\partial t} \tag{11.54}$$

Expressions (11.49) through (11.54) all have the form of "diffusion equations", well-known from the theory of heat flow (c.f. Carslaw and Jaeger, 1959).

Horizontal Flow Equation in Terms of the Storativity and Transmissivity of a Confined Aquifer

Definition: The *storativity* or *storage coefficient* of a confined aquifer is the differential volume of stored water *per unit surface area* released from (or absorbed by) an aquifer per unit decrease (or increase) of hydraulic head.

The storativity of a confined aquifer of thickness b is given by

$$S = b\,S_S \tag{11.55}$$

or, upon substituting for S_S from (11.43) above, (11.55) can be rewritten as

$$S = [\alpha + n\beta]\,\rho g b \tag{11.56}$$

The storativity is dimensionless, and typical values for confined aquifers are .005 to .00005 (Freeze and Cherry, 1979).

The *transmissivity* T of a confined aquifer is given by

$$T = b\,K \tag{11.57}$$

Multiplying the left and right sides of (11.50) by b, we have for a heterogeneous isotropic medium

$$\frac{\partial}{\partial x}\left(T\frac{\partial h}{\partial x}\right) + \frac{\partial}{\partial y}\left(T\frac{\partial h}{\partial y}\right) = S\frac{\partial h}{\partial t} \tag{11.58}$$

Or, for a homogeneous isotropic medium, factoring the transmissivity T to the denominator of the right hand side, leads to

$$\frac{\partial^2 h}{\partial x^2} + \frac{\partial^2 h}{\partial y^2} = \frac{S}{T}\frac{\partial h}{\partial t} \tag{11.59}$$

where S is the storativity and T is the transmissivity. In one-dimension, (11.59) reduces to

$$\frac{\partial^2 h}{\partial x^2} = \frac{S}{T}\frac{\partial h}{\partial t} \tag{11.60}$$

UNCONFINED AQUIFERS: MECHANISMS OF WATER RELEASE

General Flow Relation

We noted previously that the general flow relation developed above as

$$\left[\frac{\partial}{\partial x}\left(K\frac{\partial h}{\partial x}\right) + \frac{\partial}{\partial y}\left(K\frac{\partial h}{\partial y}\right) + \frac{\partial}{\partial z}\left(K\frac{\partial h}{\partial z}\right)\right] = \frac{\partial S_w}{\partial t} \tag{11.48}$$

applies to both confined and unconfined aquifers. In the case of unconfined aquifers, however, the transient storage term on the right hand side represents *two* types of sources or sinks related to dynamic changes in hydraulic head which are discussed, respectively, in the following two sections.

Specific Storage

The first type of transient storage mechanism is the release (or take up) of fluid by local elastic deformation; precisely the same process that applied to the confined aquifer case. This contribution was accounted for earlier in this chapter by the the *specific storage* , and for convenience is repeated here as

$$S_s = [\alpha + n\beta] \rho g \qquad (11.43)$$

where α is the compressibility of the host matrix, β is the compressibility of the fluid, n is the porosity of the material, ρ is the density of water, and g is the gravitational force per unit mass.

Specific Yield

Gravity drainage. In addition to the effects of local elastic deformation related to the specific storage as described above, for unconfined aquifers there is a transient storage term associated with the fact that a change in hydraulic head is related to an actual change in the elevation of the water table, hence a change in the thickness of the saturated thickness of the aquifer. This is illustrated in Figure 11.4, which shows the initial configuration of the water table on the left in Panel a). On the right, in Panel b), are shown the effects of a drawdown of an unconfined aquifer: a lowering of the water table δh over a horizontal area $\Delta A = 1$ m^2, resulting in the physical drainage of a fluid volume V_d from the top of the saturated section.

A parameter used to describe this process is the *specific yield* — defined in Chapter 1 as the ratio

$$S_y = V_d / V_t \qquad (11.61)$$

where V_d is the volume of water released under gravity drainage, and V_t is the total volume of the initially saturated material.

Horizontal flow in unconfined aquifers. It is computationally convenient, when discussing horizontal flow in unconfined aquifers, to describe the contribution of S_y to the transient storage of the *entire* thickness of the aquifer. To do so, consider first a vertical rectilinear prism having a saturated thickness h and a horizontal cross section ΔA, for a total volume

$$V_t = h \Delta A \qquad (11.62)$$

Clearly, a decrease in the hydraulic head of δh will be related to a drained volume of

$$V_d = \delta h \Delta A \qquad (11.63)$$

so that, from (11.61)

$$S_y = \delta h \Delta A / h \Delta A \qquad (11.64)$$

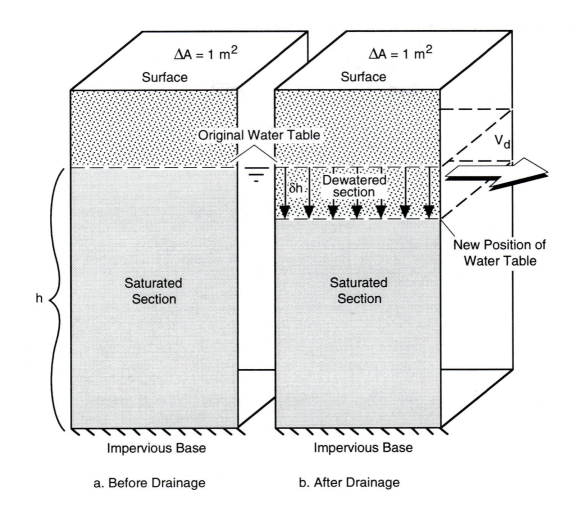

Figure 11.4 Gravity drainage of an unconfined aquifer during drawdown. **Panel a:** Initial configuration of the water table. **Panel b:** Lowering the water table δh over a horizontal area $\Delta A = 1 \text{ m}^2$ results in the physical drainage of a fluid volume V_d from the top of the saturated section.

In other words, it is convenient in the present context to define the *specific yield* S_y as the amount of water released from storage for a unit horizontal surface area through gravity drainage over the entire thickness of the aquifer for a unit decrease in the height of the water table. According to (11.61) or (11.64), the specific yield is in units of m^3 of water released per m^3 of aquifer under a footprint of 1 m^2, which of course is dimensionless.

The *storativity of an unconfined aquifer* is a measure of *both* of these contributions:

$$S = S_y + hS_s \qquad (11.65)$$

where h is the local thickness of the saturated section (i.e. the elevation of the water table relative to the base of the aquifer).

Definition: The storativity of an unconfined aquifer (11.68) is defined as the amount of water released from storage for a unit horizontal surface area through the combination of gravity drainage and elastic deformation over the entire thickness of the aquifer for a unit decrease in the height of the water table. The storativity is dimensionless.

Comparing Storage Characteristics of Confined and Unconfined Aquifers

Typical values of aquifer storage characteristics are the following.

- The specific storage S_s of water-saturated, unconsolidated materials ranges from 3×10^{-4} to 3×10^{-9} m^{-1} (Roscoe Moss Company, 1990). Storativities [S is dimensionless] for confined aquifers are typically from .005 to .00005 (Freeze and Cherry, 1979).

- Specific yields [S_y] for unconfined aquifers are 0.01 to 0.3.

The conventional paradigm is that the storativities of *unconfined* aquifers are generally much higher than the storativities of *confined* aquifers. This is due to the fact that an unconfined aquifer is actually *dewatered* during drawdown so that the specific yield from gravity drainage often dominates the total water released. In such a case, S_y will usually be much larger than hS_s, so that the storativity of an unconfined aquifer is assumed to be dominated by its specific yield. Some authors even neglect the contribution of the elastic component — the specific storage S_s — and refer to the specific yield alone as the *unconfined storativity*. However, if h is large, and S_y tends to be on the lower side of its possible range, then the depth integrated specific storage term, hS_s, can be significant, or even dominant in the overall storativity of an unconfined aquifer.

UNCONFINED AQUIFERS: THE TRANSIENT FLOW EQUATION

Conservation Condition for Transient Horizontal Unconfined Flow

We will assume that the Dupuit conditions pertain to horizontal flow in an isotropic medium. As in Chapter 8, we define a horizontal total discharge vector

$$\mathbf{Q} = [Q_x(x,y), Q_y(x,y)] \qquad (11.66)$$

Each component of the total discharge vector is the product of the corresponding component of the *specific discharge* vector and the *thickness* of the saturated section of the aquifer h. Thus, the discharge vector components are

$$Q_x = h \, q_x \qquad (11.67)$$

and

$$Q_y = h \, q_y \qquad (11.68)$$

Substituting for q_x and q_y from Darcy's law, we obtain

$$Q_x = - K h \frac{\partial h}{\partial x} \qquad (11.69)$$

and

$$Q_y = - K h \frac{\partial h}{\partial y} \qquad (11.70)$$

Following the discussion in Chapter 8, it should be a straight forward matter to convince oneself that in the limit of infinitesimal volumes the left hand side of the conservation relation (11.2) for horizontal flow reduces to the horizontal divergence of the total discharge vector $\mathbf{Q}$, and the source term on the right hand side of (11.2) is the depth integrated transient storage. Thus in the infinitesimal limit, (11.2) becomes

$$\frac{\partial}{\partial x}[Q_x] + \frac{\partial}{\partial y}[Q_y] = -\int_0^h \frac{\partial S_w}{\partial t} \, dz \qquad (11.71)$$

Next we note that the integral on the right hand side of the latter expression can be reordered and equated to the total transient storage term throughout the vertical saturated section of the aquifer, thus becoming

$$-\frac{\partial}{\partial t}\left[\int_0^h S_w \, dz\right] = -\left[S_y + h S_s \right]\frac{\partial h}{\partial t} \qquad (11.72)$$

where S_y is the specific yield and S_s is the specific storage. Upon using the definition in (11.65)

$$-\frac{\partial}{\partial t}\left[\int_0^h S_w \, dz\right] = -S \frac{\partial h}{\partial t} \qquad (11.73)$$

where S is the storativity. Thus, (11.71) becomes

$$\frac{\partial}{\partial x}[Q_x] + \frac{\partial}{\partial y}[Q_y] = -S \frac{\partial h}{\partial t} \qquad (11.74)$$

which is the conservation condition in differential form for horizontal unconfined flow under the Dupuit assumptions, S being the storativity defined by (11.65).

Transient Horizontal Flow Relation for Unconfined Aquifers

Substituting the right hand side of (11.69) and (11.70) from Darcy's law, respectively, for Q_x and Q_y in (11.74), we obtain

$$\frac{\partial}{\partial x}\left(Kh \frac{\partial h}{\partial x} \right) + \frac{\partial}{\partial y}\left(Kh \frac{\partial h}{\partial y} \right) = S \frac{\partial h}{\partial t} \qquad (11.75)$$

This is the fundamental transient flow relation for the hydraulic head in a heterogeneous, isotropic unconfined aquifer. For one dimensional flow, (11.75) reduces to

$$\frac{\partial}{\partial x}\left(Kh \frac{\partial h}{\partial x} \right) = S \frac{\partial h}{\partial t} \qquad (11.76)$$

Assuming the medium is homogeneous and isotropic, the flow relation (11.75) becomes

$$\frac{\partial}{\partial x}\left(h \frac{\partial h}{\partial x} \right) + \frac{\partial}{\partial y}\left(h \frac{\partial h}{\partial y} \right) = \frac{S}{K} \frac{\partial h}{\partial t} \qquad (11.77)$$

For one dimensional flow, (11.77) reduces to

$$\frac{\partial}{\partial x}\left(h \frac{\partial h}{\partial x} \right) = \frac{S}{K} \frac{\partial h}{\partial t} \tag{11.78}$$

While, for example, relations (11.75) and (11.77) for *unconfined* flow are analogous to (11.58) and (11.59) for *confined* flow, there are several significant differences. First the equations for unconfined flow are nonlinear second order differential equations in h which need to be "linearized" in order to employ the usual methods of differential equations to solve them. Second, the storativity for unconfined aquifers has two contributions from two distinctly different physical mechanisms, each of which operates on its own time scale. Moreover, each of these two mechanisms involves different flow paths over which the "released" fluid gets into the system. The gravity drainage term S_y for example, often leads to a "*delayed yield*" effect that modifies the drawdown of an unconfined aquifer over longer time scales. While we briefly touch upon these issues later in the text, the reader may wish to refer to more detailed literature to see how they are dealt with in theory and current practice. Several primary references are Boulton (1954), Neuman and Witherspoon (1969) and Neuman (1972); with recent overviews of current practice given by de Marsily (1986), Roscoe Moss Company (1990) and Fetter (1994).

A Simple Linearization of the Horizontal Unconfined Flow Relation

While we consider a more elegant approach for linearizing the unconfined flow relation later in the text, here we introduce the concept using a simple, though restrictive, argument. Assume that the instantaneous value of hydraulic head can be represented by a pre-event static level (or a constant average level) h_o plus a perturbation term δh given by

$$\delta h = h(t) - h_o \tag{11.79}$$

so that

$$h(t) = h_o + \delta h \tag{11.80}$$

Introducing (11.80) into (11.77) for example, leads to

$$\frac{\partial}{\partial x}\left(\left(h_o + \delta h \right) \frac{\partial\, \delta h}{\partial x} \right) + \frac{\partial}{\partial y}\left(\left(h_o + \delta h \right) \frac{\partial\, \delta h}{\partial y} \right) = \frac{S}{K} \frac{\partial\, \delta h}{\partial t} \tag{11.81}$$

where we have used the fact that derivatives of h_o are zero. Upon differentiating further, we obtain

$$\left[\frac{\partial}{\partial x}\left(h_o + \delta h \right) \right] \frac{\partial\, \delta h}{\partial x} + \left(h_o + \delta h \right) \frac{\partial}{\partial x}\left(\frac{\partial\, \delta h}{\partial x} \right)$$

$$+ \left[\frac{\partial}{\partial y}\left(h_o + \delta h \right) \right] \frac{\partial\, \delta h}{\partial y} + \left(h_o + \delta h \right) \frac{\partial}{\partial y}\left(\frac{\partial\, \delta h}{\partial y} \right) = \frac{S}{K} \frac{\partial\, \delta h}{\partial t} \tag{11.82}$$

and collecting terms,

$$\left(\frac{\partial\, \delta h}{\partial x} \right)^2 + \left(h_o + \delta h \right)\left(\frac{\partial^2\, \delta h}{\partial x^2} \right)$$

$$+ \left(\frac{\partial\, \delta h}{\partial y} \right)^2 + \left(h_o + \delta h \right)\left(\frac{\partial^2\, \delta h}{\partial y^2} \right) = \frac{S}{K} \frac{\partial\, \delta h}{\partial t} \tag{11.83}$$

- 160 -

Keeping only those terms on the left hand side multiplied by h_o, and differentiating terms in δh, we obtain

$$h_o \frac{\partial^2 h}{\partial x^2} + h_o \frac{\partial^2 h}{\partial y^2} = \frac{S}{K} \frac{\partial h}{\partial t} \qquad (11.84)$$

Earlier in this chapter we defined the transmissivity of a confined aquifer by

$$T = b\,K \qquad (11.57)$$

where b is the thickness of the confined layer and b is its hydraulic conductivity. We want to define an analogous parameter here for an unconfined aquifer. Noting that, for an unconfined aquifer, h_o is its unperturbed saturated thickness. For small δh, we define the transmissivity of an unconfined aquifer by

$$T = h_o\,K \qquad (11.85)$$

Thus, upon factoring h_o on the left hand side of (11.84) to the denominator of its right hand side, and using (11.85), we obtain the following form for horizontal unconfined flow

$$\frac{\partial^2 h}{\partial x^2} + \frac{\partial^2 h}{\partial y^2} = \frac{S}{T} \frac{\partial h}{\partial t} \qquad (11.86)$$

which is exactly the same form as (11.59) for *confined* flow. We conclude, therefore, that solutions to the horizontal flow equation (11.86) apply to both confined aquifers and unconfined aquifers, providing that, in the latter case, $\delta h \ll h_o$, which is to say that perturbations in h have to be much smaller than the saturated thickness of the unconfined layer. While this condition is quite restrictive on the kinds of problems that can be treated using this paradigm, it significantly simplifies the task of solving problems to which it *can* be applied, and is often used in situations where the assumption of δh being small might not pertain, but where even an approximate solution might provide useful insight with a minimum of computational investment.

Chapter 12. Transient 1-D Flow: Elementary Periodic and Aperiodic Forcing Terms

CONCEPT OF ELEMENTARY SOLUTIONS OF THE TRANSIENT FLOW EQUATION

This chapter describes solutions to the 1-D transient flow equation for several simple classes of periodic and aperiodic forcing terms. These solutions are as fundamental to representing the dynamics of groundwater flow as are sine and cosine functions to representing the concepts of trigonometry, or as exponents, logarithms and power series are to representing the concepts of algebra. The underlying theme of this and the next chapter is that, having determined one or more sets of what various authors refer to as "elementary", "fundamental" or "primitive" solutions for relatively simple periodic or aperiodic forcing terms, one can then synthesize or combine these solutions to solve problems involving more complicated source distributions and/or histories. In this chapter we derive several of the underpinning elementary solutions for 1-D transient flow. We first consider the case of an aquifer driven by a periodic, simple harmonic variation in hydraulic head; we next consider the case of an aquifer subjected to a local impulsive change in hydraulic head. In the next chapter we combine or synthesize such solutions to simulate the response of a system to more complicated sources.

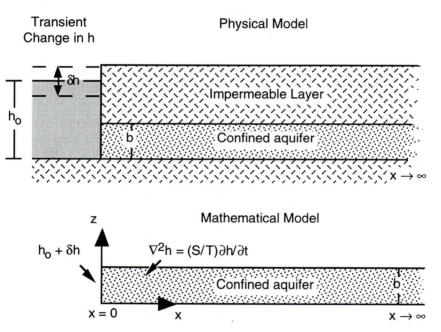

Figure 12.1. A 1-D semi-infinite confined aquifer with a transient source at x = 0.

We begin by recalling that one dimensional transient flow in the horizontal x direction in a confined aquifer can be represented by the diffusion equation

$$\frac{\partial^2 h}{\partial x^2} = \frac{S}{T}\frac{\partial h}{\partial t} \tag{12.1}$$

where S is the storativity and T is the transmissivity. Of the many types of situations that fall into this class of models, one example is shown in Figure 12.1 which shows a confined aquifer extending from x = 0 on the left to x = infinity on the right.

For the case shown here, the hydraulic head in the aquifer is driven by a constant head, h_o, at x = 0, on which is superimposed a time variation $\delta h(t)$. An objective of solving this class of problems might be to predict the temporal and spatial behavior of the hydraulic head within the aquifer due to a given time variation $\delta h(t)$ at its boundary x = 0. As discussed in the last chapter, such a situation can also approximate the case of an *unconfined* aquifer, providing $\delta h(t) \ll h_o$; which is to say providing one can neglect variations in the saturated thickness of the aquifer due to time variations of the hydraulic head.

For the purposes of the present discussion, however, we will assume that we are dealing with a *confined* aquifer, and that the time dependence of the transient phenomena can be described either as a periodic (cyclic) function in time or as an aperiodic (noncyclic) function in time. We will first consider the case of a *periodic* transient.

PERIODIC TRANSIENT FLOW IN A CONFINED AQUIFER

Method of Separation of Variables

We assume that the hydraulic head can be expressed as the product of two functions

$$h(x,t) = X(x)\ T(t) \qquad (12.2)$$

where $X(x)$ is a function only of x and $T(t)$ is a function only of t.

Note: The use of $T(t)$ to represent the function of time follows standard mathematical practice, but the reader will need to exercise some caution in not confusing the notation with T used for the transmissivity. In this text the time function will always be associated with an explicit time dependence shown by brackets, e.g. "T(t)". This notation is favored, as is "X(x)" to denote the function of x, in preference to employing other arbitrary letters of the alphabet.

Substituting (12.2) into (12.1), and performing the necessary differentiations, the result can be arranged in the form

$$\frac{1}{X(x)}\frac{\partial^2 X(x)}{\partial x^2} = \frac{S}{T}\frac{1}{T(t)}\frac{\partial T(t)}{\partial t} \qquad (12.3)$$

Since the left hand side is a relation only in x and the right hand side is a relation only in t, the individual terms on each side must vary so that each side is a constant.

We set this constant equal to k^2, so that

$$\frac{d^2 X(x)}{dx^2} = k^2\ X(x) \qquad (12.4)$$

and

$$\frac{dT(t)}{dt} = k^2\frac{T}{S}\ T(t) \qquad (12.5)$$

which are two ordinary differential equations with a common separation constant k^2.

Harmonic Time Dependence: Fourier Integral Transforms

The temporal behavior of any but the most pathologic functions can usually be decomposed into a set of frequency domain Fourier harmonics. Functions which are periodic in time are best represented as Fourier *series*, while aperiodic transient functions are best represented as Fourier *integral transforms*.

For example, the temporal function T(t) in (12.5) can be represented using the following Fourier transform pair:

$$T(t) = \frac{1}{2\pi} \int_{-\infty}^{+\infty} F(\omega) \, e^{i\omega t} \, d\omega \qquad (12.6)$$

where ω is the radian frequency and $F(\omega)$ is a complex function of ω given by

$$F(\omega) = \int_{-\infty}^{+\infty} T(t) \, e^{-i\omega t} \, dt \qquad (12.7)$$

Note that ω has units of radians per unit time, and $\omega = 2\pi f$, where f is the periodic frequency in cycles per unit time. $F(\omega)$ is often referred to as the frequency domain Fourier spectrum.

We allow T(t) to be represented by one of the harmonic terms in (12.6)

$$T(t) = F(\omega) \, e^{i\omega t} \qquad (12.8)$$

Substituting (12.8) into (12.5), we find, after the appropriate differentiation on the left hand side, and substituting T(t) for $F(\omega) \, e^{i\omega t}$, that

$$i\omega \, T(t) = k^2 \frac{T}{S} T(t) \qquad (12.9)$$

Cancelling the common factor T(t), we can solve for the following condition on the separation constant k^2

$$k^2 = i\omega \frac{S}{T} \qquad (12.10)$$

We next solve the second order ordinary differential equation (12.3) for the functional form of X(x). By inspecting (12.3), we are tempted to try an exponential solution of the form

$$X(x) \propto e^{\pm kx} \qquad (12.11)$$

Upon substituting (12.11) into (12.3), and differentiating twice on the left hand side, we find

$$k^2 e^{\pm kx} = k^2 e^{\pm kx} \tag{12.12}$$

thus proving that (12.11) is a solution of (12.3), where

$$k = \sqrt{i\omega \frac{S}{T}} \tag{12.13}$$

It is well known that

$$\sqrt{i} = \frac{1+i}{\sqrt{2}} \tag{12.14}$$

Upon substituting (12.14) into (12.13), k can be explicitly expressed as a complex variable

$$k = \sqrt{\frac{\omega S}{2T}} + i\sqrt{\frac{\omega S}{2T}} \tag{12.15}$$

The parameter $[(\omega S)/(2T)]^{1/2}$ is so often used in hydrologic studies that we will denote it by

$$\gamma = \sqrt{\frac{\omega S}{2T}} \tag{12.16}$$

so that

$$k = \gamma + i\gamma \tag{12.17}$$

recalling that, in general, k can be either positive or negative.

At this point, it is appropriate to appraise our progress. Through the method of separation of variables, we have found that the product solution given by (12.2), $h(x,t) = X(x) T(t)$, leads to a harmonic solution to the diffusion equation given by (12.6). The $T(t)$ term can now be expressed through the relation (12.8), and the $X(x)$ term through the relation (12.11), where the separation constant k is given by (12.17), and the parameter γ is given by (12.16). Combining these results, our solution to the 1-D diffusion equation (12.1) can thus be expressed in the general form

$$h(x,t) = A(\omega) e^{\gamma x} e^{i(\omega t + \gamma x)} + B(\omega) e^{-\gamma x} e^{i(\omega t - \gamma x)} \tag{12.18}$$

where, for now, we have suppressed the constant static head term h_0 to which (12.18) would usually be added. The first term on the right side of (12.18), with the coefficient $A(\omega)$, represents a wave traveling in the negative x direction, which is exponentially attenuated with distance, because of the $e^{\gamma x}$ factor, as it propagates toward negative x.

The second term on the right, with the coefficient $B(\omega)$, represents a wave traveling in the positive x direction, which is exponentially attenuated with distance, because of the $e^{-\gamma x}$ factor, as it propagates toward positive x.

For a periodic disturbance in hydraulic head at $x = 0$, propagating in the direction of increasing x, (12.18) reduces to the form

$$h(x,t) = B(\omega) e^{-\gamma x} e^{i(\omega t - \gamma x)} \tag{12.19}$$

Such periodic variations might be due to tidal effects, waves, seasonal or climatic variations, or to a variety of cyclic effects from domestic, industrial or agricultural practices. The physical phenomenon represented by (12.19) is actually a "wave-like" disturbance of the hydraulic head that travels in the positive x direction, and is damped or attenuated with increasing distance.

An example of such a disturbance, at the period of a tidal cycle, is shown for various times in Figure 12.2.

Characteristic Attenuation Length

Each wave travels a characteristic distance $x_c = 1/\gamma$ before it attenuates to $1/e$ of its initial value. We will call this distance the "characteristic attenuation length"

$$x_c = \frac{1}{\gamma} \tag{12.20}$$

or, upon substituting (12.16),

$$x_c = \sqrt{\frac{2T}{\omega S}} \tag{12.21}$$

Note: The *characteristic attenuation length* is analogous to the *depth of penetration* or *skin depth* from the theory of the propagation of electromagnetic waves in conductive media.

Clearly, from inspecting (12.21) it can be seen that if the transmissivity of an aquifer is large, x_c will be large. The aquifer is relatively permeable and open, and the "wave" will propagate with little attenuation. On the other hand, if the storativity is large, the aquifer will readily "absorb" a disturbance, and x_c will be correspondingly small. Moreover, if the hydraulic head fluctuates rapidly (i.e. if the driving source has a high characteristic frequency ω), x_c will be small, and waves will also attenuate rapidly with distance.

Phase Velocity

The complex exponential term

$$e^{i(\omega t - \gamma x)} \tag{12.22}$$

represents a harmonic function travelling toward increasing x. This is because, for a given value of the argument

$$\omega t - \gamma x = \theta_c \tag{12.23}$$

x has to progressively increase as time increases in order for (12.23) to remain constant. Suppose one wanted to move along the aquifer in space and time (x and t) such that they observed a constant amplitude of the cyclic component of the wave, say its minimum or maximum value. If one stood still, the hydraulic head would fluctuate; but if one moved in such a way that their x position increased as t inexorably increased, they could keep the phase, θ_c, constant, and could keep up with the "wave".

How fast does x have to increase to maintain θ_c constant? This is the "phase velocity" of the wave which is determined by differentiating (12.23) with respect to time.

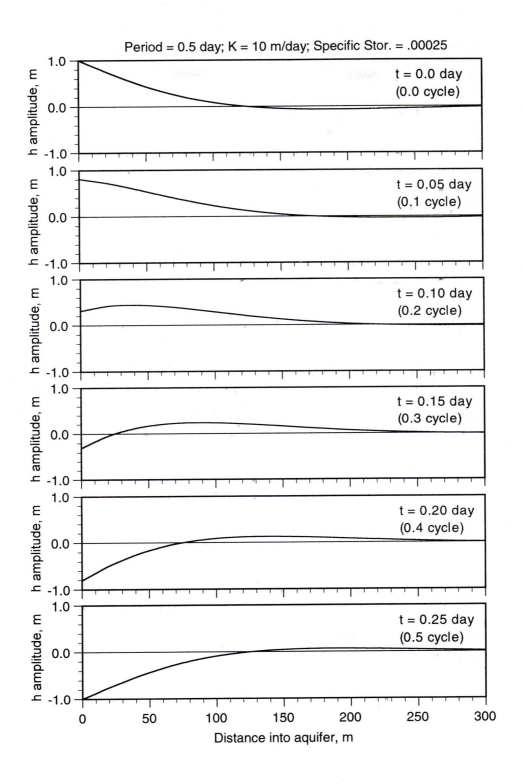

Figure 12.2 Perturbation in hydraulic head as a function of distance into a confined aquifer at various times during a tidal cycle (period = 0.5 days). The first half cycle of the event is shown for hydraulic parameters of K = 10 m day^{-1} and S_s = 0.00025.

Since θ_c is constant, we would have $\partial\theta_c/\partial t = 0$, and

$$\omega - \gamma\frac{\partial x}{\partial t} = 0 \qquad (12.24)$$

Defining the phase velocity v_p as the rate of change of position x with time, we have

$$v_p = \frac{\partial x}{\partial t} \qquad (12.25)$$

which upon substituting into (12.24) leads to

$$v_p = \frac{\omega}{\gamma} \qquad (12.26)$$

or, upon substituting (12.16) for γ

$$v_p = \sqrt{\frac{2\omega T}{S}} \qquad (12.27)$$

Characteristic Delay Time

The "delay time" for a wavelet to travel a distance x is simply given by distance over time

$$t_d = \frac{x}{v_p} \qquad (12.28)$$

which upon substituting (12.27) becomes

$$t_d = x\sqrt{\frac{S}{2\omega T}} \qquad (12.29)$$

Inspecting (12.29), we see that, whereas the delay time for a transient is directly dependent on distance, it depends on the *square root* of the storativity. Moreover, the delay time is *inversely* proportional to the square root of the frequency of the fluctuation and to the square root of the transmissivity of the aquifer. A higher storativity tends to elastically absorb more of the transient effects, with a concomitant *increase* in the delay time. On the other hand, a larger transmissivity implies less restricted flow, with a concomitant *decrease* in delay time.

APERIODIC TRANSIENT FLOW IN A CONFINED AQUIFER

General Solution

In the case of *aperiodic* transient conditions in an aquifer, it is usually better to seek a solution of the diffusion equation which is a superposition of *aperiodic* basis functions, rather than a superposition of the harmonic functions as was done above for the periodic case. Such solutions are commonly required when one is studying the response of an aquifer following a sudden change in the applied hydraulic head, or recharge/discharge conditions.

We begin with the diffusion equation for the case of transient one dimensional confined flow in the horizontal x direction [(12.1) above, repeated here for convenience]:

$$\frac{\partial^2 h}{\partial x^2} = \frac{S}{T}\frac{\partial h}{\partial t} \qquad (12.1)$$

where we will assume from the outset that the transient phenomena is aperiodic (non-cyclic) in time.

Consider the problem illustrated in Figure 12.3. We have a locally prescribed function for the initial values of the hydraulic head $h(x,0) = f(x)$ over a section of a one dimensional confined aquifer. Our objective is to predict the behavior of the head at later times t for various positions x.

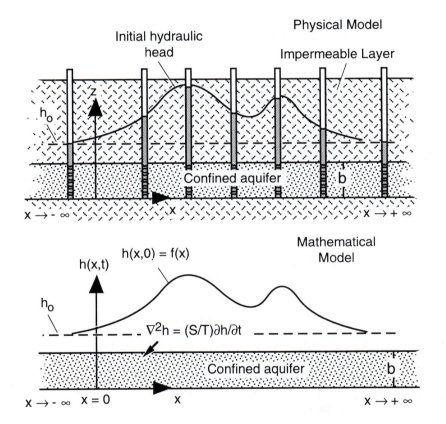

Figure 12.3 Conceptual model for representing transient saturated flow in an infinite $(-\infty < x < +\infty)$ 1-D confined aquifer of thickness b. The initial condition of the hydraulic head $h(x,0) = f(x)$ in this case is given by observation wells (also known as piezometers or potentiometers).

We assume that the spatial dependence of the hydraulic head $h(x,t)$ can be represented by the following Fourier transform

$$h(x,t) = \frac{1}{2\pi}\int_{-\infty}^{+\infty} F(k,t)\, e^{ikx}\, dk \qquad (12.30)$$

where k is the spatial wave number and $F(k,t)$ is a complex function of k and t given by

$$F(k,t) = \int_{-\infty}^{+\infty} h(x,t)\, e^{-ikx}\, dx \qquad (12.31)$$

Of course for (12.30) and (12.31) to be integrable, the functional forms for $h(x,t)$ and $F(k,t)$ should be sufficiently "well behaved" in an analytical sense.

Arranging (12.1) in the form

$$\frac{\partial^2 h}{\partial x^2} - \frac{S}{T}\frac{\partial h}{\partial t} = 0 \qquad (12.32)$$

we substitute (12.30) for $h(x,t)$, and rearrange the order of integration and differentiation (which is to bring the derivative operations within the integration) to obtain

$$\frac{1}{2\pi}\int_{-\infty}^{+\infty}\left[F(k,t)\left(-k^2\right)e^{ikx} - \frac{S}{T}\frac{\partial F(k,t)}{\partial t}e^{ikx} \right]dk = 0 \qquad (12.33)$$

The reader will note that the first term in the square bracket is the consequence of the $\partial^2 h/\partial x^2$ operation implied in (12.32), whereas the second term in the square bracket of (12.33) is the consequence of the $(S/T)\partial h/\partial t$ operation in (12.32). Rearranging the terms in the square bracket to the form

$$\frac{1}{2\pi}\int_{-\infty}^{+\infty}\left[F(k,t)\left(-k^2\right) - \frac{S}{T}\frac{\partial F(k,t)}{\partial t} \right]e^{ikx}\, dk = 0 \qquad (12.34)$$

we note that the only way for the integral in (12.34) to be equal to zero is for the term in the square bracket to be equal to zero, or

$$\frac{\partial F(k,t)}{\partial t} = -k^2\left(\frac{T}{S}\right)F(k,t) \qquad (12.35)$$

The latter relation clearly has the solution

$$F(k,t) = F(k,0)\exp\left(-k^2\left(\frac{T}{S}\right)t\right) \qquad (12.36)$$

where here and elsewhere in this text the notation "$\exp(u)$" will refer to the *exponential* of u to the base e.

The wave number spectrum term $F(k,0)$ on the right hand side of (12.36) is obtained from the inverse Fourier transform of the initial condition for the hydraulic head at $t = 0$ given by $f(x)$. Thus, according to (12.31) we have

$$F(k,0) = \int_{-\infty}^{+\infty} f(x)\, e^{-ikx}\, dx \qquad (12.37)$$

which completes the formal solution to the problem that we posed with reference to Figure 12.3.

To summarize the above procedure, the initial condition f(x) is substituted into (12.37) leading to the time invariant wave number spectrum F(k,0). Substituting F(k,0) into (12.36), we obtain the time dependent wave number spectrum F(k,t). Substituting the latter into the Fourier transform (12.30), we can synthesis the behavior of the hydraulic head h(x,t) at all positions x, for all times $t \geq 0$.

Impulse Response of an Aquifer

Properties of an impulse. The above considerations can be applied to the case of an *impulse* change in the hydraulic head from an initial steady value. An impulse is defined as a sudden increase (or decrease) in the value of the hydraulic head in the vicinity of a point, followed by a rapid return to its initial value. One way to create such an impulse is by means of a sudden influx of water at a point. This would involve a finite flux of water is delivered over a short period of time Δt within an infinitesimal vicinity ($\pm \Delta x/2$) of the point x_0. We do not care so much about the detailed functional form of this impulse, just that the total volume of fluid delivered has a specified value, that the duration of the input (or output) flow be short relative to the principal time constant(s) of the system, and the spatial scale over which it occurs be small compared to the principal dimension(s) of the system.

Three types of "impulse functions" are depicted in Figure 12.4. Each is scaled so that they have unit area under their respective curves.

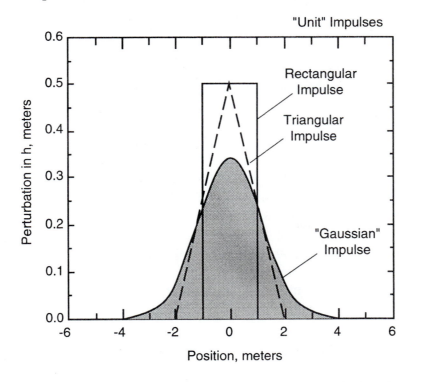

Figure 12.4. Examples of three classes of unit impulses: A rectangular impulse; a triangular impulse; a "Gaussian" impulse. All have the same unit area under their respective curve over the 1-D interval ($-\infty < x < +\infty$).

The area under the *rectangular* impulse is unity providing its amplitude is related to its base Δx by

$$\text{Ampl (rectangle)} = 1.0 \, / \, \Delta x \qquad (12.38)$$

Similarly, the area under the *triangular* impulse is unity providing its amplitude is related to its base Δx by

$$\text{Ampl (triangle)} = 1.0 \, / \, (2\Delta x) \qquad (12.39)$$

Finally, it is well known that the area under the "Gaussian" impulse is unity providing its amplitude, or the coefficient $a \cdot (\pi)^{-1/2}$, is related to its argument such that

$$\frac{a}{\pi^{1/2}} \int_{-\infty}^{+\infty} \exp\left(-a^2 x^2\right) dx = 1.0 \qquad (12.40)$$

The Dirac delta function. The concept of the "impulse function" as discussed above is related to that of the *Dirac delta function* from mathematical physics. For an event at the origin, the latter is defined by

$$\delta(x) = 0 \qquad x \neq 0 \qquad (12.41a)$$

$$\lim_{x \to 0} \delta(x) = \infty \qquad (12.41b)$$

$$\int_{-\infty}^{+\infty} \delta(x) \, dx = 1 \qquad (12.41c)$$

Forms for the Dirac delta function can be defined for any point x_o along the x axis, according to the set of relations (12.41) with the origin shifted to x_o. In such a case the delta function is denoted by $\delta(x-x_o)$.

Impulses as delta functions. The parallels between the Dirac delta function and the functions sketched in Figure 12.4 should be clear. For example, let us denote the rectangular impulse with the symbol $\delta_{rect}(x)$ to emphasize this connection. That $\delta_{rect}(x)$ is a good approximation to the Dirac delta function can be seen from its definition

$$\delta_{rect}(x) = 1/\Delta x \quad \text{for} \quad -\Delta x/2 \leq x \leq +\Delta x/2 \qquad (12.42)$$

resulting in the following conditions analogous to those summarized by (12.41a) through (12.41c) for the Dirac delta function:

$$\delta_{rect}(x) = 0 \quad \text{for} \quad x < -\Delta x/2 \text{ or } x > +\Delta x/2 \qquad (12.43a)$$

$$\lim_{\Delta x \to 0} \delta_{rect}(x) = \lim_{\Delta x \to 0} 1/\Delta x = \infty \qquad (12.43b)$$

$$\int_{-\infty}^{+\infty} \delta_{\text{rect}}(x)\, dx = \int_{-\Delta x/2}^{+\Delta x/2} \left(\frac{1}{\Delta x} \right) dx = \frac{1}{\Delta x} \int_{-\Delta x/2}^{+\Delta x/2} dx = 1 \qquad (12.43c)$$

Response of the hydraulic head to a unit impulse forcing term. Let us now determine the response of a confined aquifer to a unit impulse in the hydraulic head. To simplify the algebra, but without loss of generality, we will assume that the impulse occurs at the origin $x = 0$, and at time $t = 0$. In such a case, the initial function $f(x)$ introduced in an earlier section will be restricted to non-zero values in a small interval about the origin, and for times close to $t = 0$. To be specific, any of the examples depicted in Figure 12.4 would satisfy this requirement. In any practical sense it does not matter whether one defines the impulse function as the Dirac delta function, one of the functions illustrated in Figure 12.4, or some other alternative to represent the impulse forcing term — although there may be some analytical or computational advantage in selecting one over the other. The important thing is that the following condition pertains to this particular form of $f(x)$:

$$\int_{-\infty}^{+\infty} f(x)\, dx = 1 \qquad (12.44a)$$

$$f(x) = 0 \qquad \text{for } x < -\Delta x/2 \text{ and } x > +\Delta x/2 \qquad (12.44b)$$

The latter condition implies that $f(x)$ is non-zero only in a small interval about the origin. As argued below, these two conditions imply that the integral in (12.37) becomes

$$F(k,0) = \int_{-\infty}^{+\infty} f(x)\, e^{-ikx}\, dx = 1 \qquad (12.45)$$

To see this latter equality, in a practical sense the interval over which $f(x)$ is non-zero is expected to be small compared to the shortest wavelength, λ, of the spatially harmonic term $\exp(-ikx)$ having the highest wavenumber ($k = 2\pi/\lambda$) in (12.45). In such a case, at $x \approx 0$, $\exp(-ikx) = \cos(kx) - i\sin(kx)$ can be assumed to be essentially unity over the interval Δx, so that the last equality in (12.45) reduces to (12.44). Strictly speaking from a *mathematical* point of view, the wavenumber k in (12.45) goes to infinity, so the proper choice for $f(x)$ should be the Dirac delta function defined by equation set (12.41). One then assumes that, in the limit of $k \to \infty$, $f(x) = \delta(x)$ is always substantially *narrower* than the highest wavenumber term $\exp(-ikx)$ in (12.45), so that (12.45) applies. For most practical problems, however, particularly for those stemming from the diffusion-like equations of groundwater flow, large wavenumber terms — even if initially present in $h(x,0)$ — attenuate very quickly with distance from the source, so that truly *infinite* wavenumbers are not of concern, and the condition of (12.45) can be achieved with any of the unit impulse functions illustrated in Figure 12.4.

Thus, assuming (12.45) applies, $F(k,0)$ from the latter expression can be substituted into (12.36), and thence $F(k,t)$ into (12.30), leading to the following form for the impulse response of the hydraulic head

$$\delta_h(x,t) = \frac{1}{2\pi} \int_{-\infty}^{+\infty} \left[\exp\left(-k^2 \left(\frac{T}{S} \right) t \right) \exp\left(-ikx \right) \right] dk \qquad (12.46)$$

where again we use the delta notation, $\delta_h(x,t)$, to emphasize the delta function behavior of $h(x,t)$ in the present situation. We note further that the time dependent spatial wave number spectrum in (12.46) is symmetric (because of the k^2 term) about $k = 0$. Thus we need to consider only even terms in the harmonic function, and (12.46) reduces to

$$\delta_h(x,t) = \frac{1}{\pi} \int_0^{+\infty} \left[\exp\left(-k^2 \left(\frac{T}{S}\right) t \right) \cos(kx) \right] dk \qquad (12.47)$$

The integral in (12.47) is now in a form that is listed in most mathematical tables of integrals. A typical identity might be

$$\int_0^{+\infty} \left[\exp\left(-a^2 u^2 \right) \cos(bu) \right] du = \frac{\sqrt{\pi}}{2a} \exp\left(-\frac{b^2}{4a^2} \right) \qquad (12.48)$$

Upon making the appropriate substitutions and associations from (12.48), the integral in (12.47) reduces to

$$\delta_h(x,t) = \frac{1}{\sqrt{\pi}} \sqrt{\frac{S}{4tT}} \exp\left(-\frac{x^2 S}{4tT} \right) \qquad (12.49)$$

which is the impulse response of the system to a unit impulse forcing term in h. This is one of the cornerstone relations in groundwater flow.

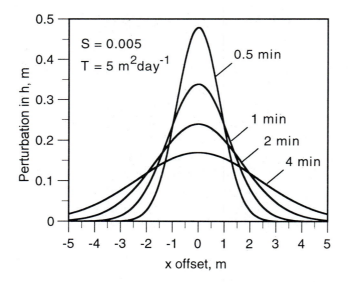

Figure 12.5 Impulse response of a horizontally infinite 1-D confined aquifer to a unit impulse forcing term occurring at the origin at t = 0, for the indicated times. A transmissivity of T = 5.0 m^2 day^{-1} and a storativity of S = .005 is assumed.

Example of an Impulse Response in Space and Time

As an example, Figure 12.5 illustrates the response of a horizontally infinite 1-D confined aquifer to a unit impulse forcing term occurring at the origin ($x = 0$) at $t = 0$. The aquifer is assumed to have a transmissivity of $T = 5.0$ m^2 day^{-1} and a storativity of $S = .005$. The impulse response is shown as a function of offset distance x from the source at various times.

For comparison, the same situation is illustrated in Figure 12.6, but the impulse response is shown as a function of time for various offset distances x from the source.

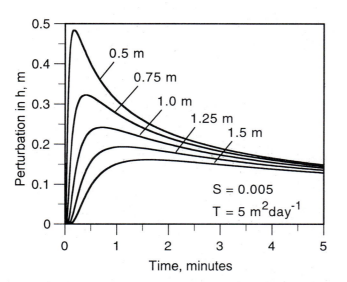

Figure 12.6. The impulse response of the same system described by Figure 12.5, but shown as a function of time for various offset distances x from the source.

Invariant Impulse Response

Multiplying both sides of (12.49) by the differential element dx, we obtain

$$\delta_h(x,t)\, dx = \frac{1}{\sqrt{\pi}}\, \sqrt{\frac{S}{4tT}}\, \exp\left(-\frac{x^2 S}{4tT}\right) dx \qquad (12.50)$$

Introducing a change of variable through the invariant parameter

$$\xi = \sqrt{\frac{x^2 S}{4tT}} \qquad (12.51)$$

(12.50) can be expressed in the form

$$\delta_h^I(\xi)\, d\xi = \frac{\xi}{\sqrt{\pi}}\, \exp\left(-\xi^2\right) d\xi \qquad (12.52)$$

where the superscript "I" denotes the function

- 175 -

$$\delta_h^I(\xi) = \frac{\xi}{\sqrt{\pi}} \exp\left(-\xi^2\right) \tag{12.53}$$

as the invariant impulse response, and has the behavior plotted in Figure 12.7.

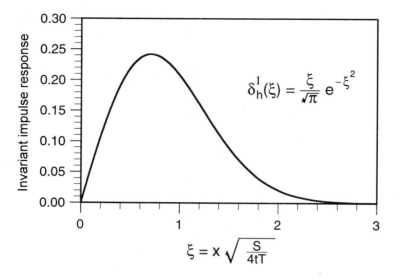

$$\delta_h^I(\xi) = \frac{\xi}{\sqrt{\pi}} e^{-\xi^2}$$

$$\xi = x \sqrt{\frac{S}{4tT}}$$

Figure 12.7. The invariant impulse parameter.

The advantage of the latter form is that for specific values of S and T, respectively, a single curve describes the dependence of the impulse response over the complete range of combinations of x and t through the parameter ξ. In addition, Figure 12.7 clearly shows that the impulse response is strongest for those combinations of x and t such that ξ is on the order of unity; and that the impulse response is significantly reduced outside the range $0.2 \le \xi \le 2.0$. This implies that hydrogeologic events for a given system undergo significant variations for combinations of x and t in the range

$$0.4 \sqrt{T/S} \le \left(x/\sqrt{t} \right) \le 4.0 \sqrt{T/S} \tag{12.54}$$

To check this, consider the case illustrated in Figure 12.5 and Figure 12.6. For the assumed values of S = .005 and T = 5 m^2 d^{-1} used to calculate these responses, we have T/S = 10^3. Thus we would expect significant variations of the impulse response to have combinations of x and t that fall with the range

$$12.6 \le \left(x/\sqrt{t} \right) \le 126 \tag{12.55}$$

From the plots of the impulse response in Figure 12.5 and Figure 12.6, typical spatial scales are on the order of 1 meter and typical time scales are 1 minute (approx. 6.9×10^{-4} day). In this case we have

$$x/\sqrt{t} = 38 \tag{12.56}$$

which not only falls with the range indicated in (12.55), but is remarkably close to the geometric mean (approximately 40.) of the two extremes.

Clearly the invariant impulse response provides useful qualitative insight into assessing the spatial and temporal scales over which hydrogeologic transients occur, and finds applications from planning and interpreting field studies to designing analytical models and numerical simulations of groundwater flow.

Chapter 13. Transient 1-D Flow: Superposition of Elementary Response Functions

COMPOSITES OF PERIODIC TRANSIENTS

Review of Harmonic Solutions

The last chapter pointed out that one class of elementary solutions to the 1-D transient flow equation

$$\frac{\partial^2 h}{\partial x^2} = \frac{S}{T} \frac{\partial h}{\partial t} \tag{12.1}$$

consists of harmonic functions of the form

$$h(x,t) = A(\omega)\, e^{+\gamma x}\, e^{+i(\omega t + \gamma x)} + B(\omega)\, e^{-\gamma x}\, e^{i(\omega t - \gamma x)} \tag{13.1}$$

where $\gamma = [(\omega S)/(2T)]^{1/2}$, ω is the radian frequency (in radians per unit time), S is the storativity and T is the transmissivity. Such functions are useful to represent the response of aquifers to *periodic* forcing terms. (**Note**: To simplify notation, here and often in the following discussion, the static hydraulic head term h_o has been suppressed on the right hand side of (13.1), but is understood to be present.)

Recall that the first term on the right side of (13.1), with the coefficient $A(\omega)$, represents a wave traveling in the *negative* x direction, and the second term on the right, with the coefficient $B(\omega)$, represents a wave traveling in the *positive* x direction. In our example of the last chapter, we considered only the latter term for the semi-infinite region $x \geq 0$, to represent a "wave" traveling in the positive x direction from a periodic forcing term at $x = 0$. However, it should be intuitive that if one or more actual sources existed to the right along the +x axis, or if one or more discontinuities were present in this region to cause reflections of the primary signal back in the negative x direction, then contributions such as the first term on the right in (13.1) would be needed to represent this exigency. Let us briefly examine this concept with an example.

Example: A Finite Length Confined Aquifer

Consider the case of a confined aquifer not extending to infinity as discussed in the last chapter, but terminating at a distance $x = L$ from a harmonic source at $x = 0$ as shown in Figure 3.1. What do we mean by a termination? For the present purpose, we assume this implies that at $x = L$ the horizontal flux q_x is zero. This in turn implies, from Darcy's law, that $\partial h/\partial x = 0$ at $x = L$. Differentiating (13.1) with respect to x, and setting the result to 0 at $x = L$ results in

$$\partial h/\partial x\big|_{x=L} = 0 = \gamma(1+i)\left[A(\omega)e^{+(1+i)\gamma L} - B(\omega)e^{-(1+i)\gamma L}\right]e^{i\omega t} \tag{13.2}$$

For the expression on the right hand side of (13.2) to be zero requires that the term in square brackets be zero, or

$$A(\omega)e^{(1+i)\gamma L} = B(\omega)e^{-(1+i)\gamma L} \tag{13.3}$$

which can be solved for

$$A(\omega) = B(\omega)e^{-2(1+i)\gamma L} \tag{13.4}$$

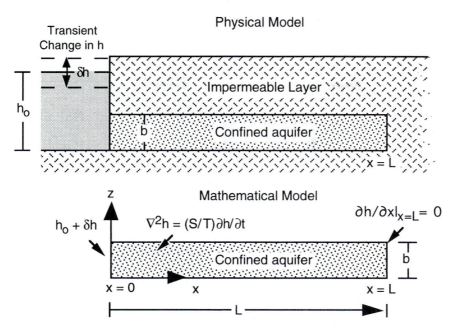

Figure 13.1 Geometry of a 1-D finite length confined aquifer, driven by a transient harmonic source at x = 0, with a condition of zero flow at x = L.

Substituting the right hand side of (13.4) for A(ω) in (13.1) we obtain

$$h(x,t) = B(\omega)\left[e^{-2(1+i)\gamma L}\, e^{\gamma x}\, e^{i(\omega t+\gamma x)} + e^{-\gamma x}\, e^{i(\omega t-\gamma x)}\right] \qquad (13.5)$$

Thus (13.5) represents the superposition of a primary wave traveling to the right (the last term in the square bracket) and a reflected wave traveling to the left (the first term in the square bracket). B(ω) is the amplitude of the hydraulic head forcing term at x = 0. The relative amplitude of the reflected wave is adjusted to match the boundary condition of zero horizontal flux at the right hand termination of the aquifer at x = L. This is a simple example of how elementary functions (in this case two harmonic functions of x and t) can be superimposed so that the composite function meets an imposed boundary condition.

COMPOSITES OF APERIODIC TRANSIENTS

Superposition of Aperiodic Transients: An Intuitive Viewpoint

In the last chapter we derived the following elementary function representing the response of a confined aquifer to a unit impulse in the hydraulic head localized at the origin

$$\delta_h(x,t) = \frac{1}{\sqrt{\pi}} \sqrt{\frac{S}{4tT}}\, \exp\left(-\frac{x^2 S}{4tT}\right) \qquad (13.6)$$

where S is the storativity and T is the transmissivity of the aquifer. Such elementary functions can be synthesized or composited to yield more general forms. Consider, for example, the case illustrated in Figure 13.2. We want to simulate the case where initially (at t = 0) the hydraulic head is locally disturbed over a finite distance by a constant value Δh_c above its static level h_o. Outside this zone of uniform disturbance, we assume the head is initially at its static value.

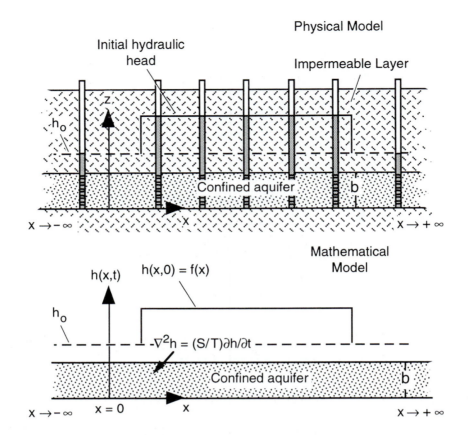

Figure 13.2 An infinite 1-D confined aquifer on which is initially imposed a uniform perturbation in the hydraulic head over a finite width.

To simplify our expressions we will shift the origin of our coordinate system to the center of this perturbed zone, such that the initial distribution of h can be described by

$$h(x,0) = h_o \qquad x < -L/2 \tag{13.7a}$$

$$h(x,0) = h_o + \Delta h_c \qquad -L/2 \leq x \leq +L/2 \tag{13.7a}$$

$$h(x,0) = h_o \qquad x > +L/2 \tag{13.7a}$$

It should be intuitive that one way to approximate this initial condition is by superimposing a set of unit impulses centered at equally spaced points over the range of $-L/2 \leq x \leq +L/2$.

The result of doing so is illustrated in Figure 13.3 for the case of 11 unit impulses distributed at intervals of 1 meter over the range from $-5.0 \leq x \leq +5.0$ meters.

This figure illustrates the individual response from each discrete event as well as the composite response of all the discrete contributions at $t = 0.5$ min. In mathematical terms, this operation is a numerical synthesis of discrete events to simulate a continuous phenomena in space (x) and time (t).

We discuss this procedure in more detail below.

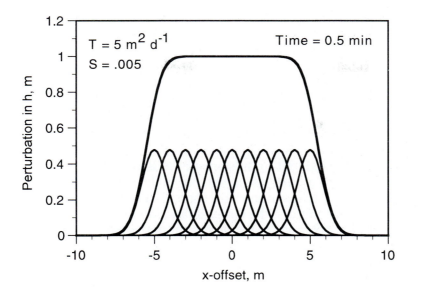

Figure 13.3 Response at t = 0.5 min from 11 individual unit impulses equally spaced from −5 to +5 meters, as well as the composite response of all the discrete contributions.

Superposition of Aperiodic Transients: A Mathematical Viewpoint

We now explore the more formal mathematical basis for the kind of superposition performed in the above section.

Mathematical preliminaries. We will be making use of the unit impulse response discussed above and in Chapter 12. Recall, for example, that one of several classes of unit impulses described was the rectangular delta function $\delta_{rect}(x)$, having an amplitude given by

$$\delta_{rect}(x) = 1/\Delta x \quad \text{for} \quad -\Delta x/2 \le x \le +\Delta x/2 \qquad (13.8)$$

and its overall behavior was defined by the relations

$$\delta_{rect}(x) = 0 \quad \text{for} \quad x < -\Delta x/2 \text{ or } x > +\Delta x/2 \qquad (13.9a)$$

$$\lim_{\Delta x \to 0} \delta_{rect}(x) = \lim_{\Delta x \to 0} 1/\Delta x = \infty \qquad (13.9b)$$

$$\int_{-\infty}^{+\infty} \delta_{rect}(x)\, dx = \int_{-\Delta x/2}^{+\Delta x/2} \left(\frac{1}{\Delta x}\right) dx = \frac{1}{\Delta x} \int_{-\Delta x/2}^{+\Delta x/2} dx = 1 \qquad (13.9c)$$

The above expressions represent a unit impulse located at x = 0. In general, a rectangular unit impulse function at some other position, say x_o, can be represented by $\delta_{rect}(x - x_o)$. The area under the curve of this function centered at x_o is given by $\delta_{rect}(x - x_o)\Delta x = 1.0$. Recall that it would be this unit area that would generate the associated impulse *response* centered at x_o:

$$\delta_h(x,t,x_o) = \frac{1}{\sqrt{\pi}} \sqrt{\frac{S}{4tT}} \exp\left(-\frac{(x-x_o)^2 S}{4tT}\right) \qquad (13.10)$$

Note that we have added another parameter, x_o, to the argument of the term on the left hand side. This is to emphasize that the particular impulse response in (13.10) represents the behavior of hydraulic head for all x $(-\infty \le x \le +\infty)$ for $t \ge 0$ due to a unit impulse at $x = x_o$.

Green's functions. This generalized unit impulse response is often referred to as a *Green's function*. Some authors refer to it as a *source function*, but in the context of the present discussion, this latter terminology might confuse the "source", a unit impulse, with the "effect", the unit impulse *response* or, if one prefers, the *Green's function* response of the system for a unit impulse forcing term. Alternative notations in the literature for the left hand side of (13.10) are equivalently

$$\delta_h(x-x_o,t); \; \delta_h(x,t,x_o); \; \delta_h(x,t;x_o); \; \delta_h(x,t\,|x_o) \qquad (13.11a)$$

or

$$G(x-x_o,t); \; G(x,t,x_o); \; G(x,t;x_o); \; G(x,t\,|x_o) \qquad (13.11b)$$

where the latter forms explicitly identify the respective expressions as "Green's functions". In all cases the expressions represent in words: "The response of the system as observed at x and t for a unit impulse forcing term at x_o."

As suggested above, in reference to the result in Figure 13.3, elementary forms such as the impulse response function can be synthesized into a composite function. This was done to construct the result in Figure 13.3 through the simple summation

$$\text{Composite } h(x,t) = \sum_{i=1}^{i=n} \left[\delta_h(x,t,x_i) \, \Delta x_i \right] \qquad (13.12)$$

where the discrete Green's function or impulse response $\delta_h(x,t,x_i)$ is given by

$$\delta_h(x,t,x_i) = \frac{1}{\sqrt{\pi}} \sqrt{\frac{S}{4tT}} \exp\left(-\frac{(x-x_i)^2 S}{4tT}\right) \qquad (13.13)$$

and where the notation x_i replaces the term x_o.

The operation in (13.12) is often referred to by various authors as a *synthesis, superposition, compositing* or *convolution*. It is useful, for more general applications, to have expressions analogous to (13.12) in the integral form that we develop below.

Superposition as an integral operation. In Chapter 12, we employed the Fourier transform pair such that the spatial dependence of the hydraulic head $h(x,t)$ can be represented by

$$h(x,t) = \frac{1}{2\pi} \int_{-\infty}^{+\infty} F(k,t) \, e^{ikx} \, dk \qquad (13.14)$$

where k is the spatial wave number, and $F(k,t)$ is a complex function of k and t given by

$$F(k,t) = \int_{-\infty}^{+\infty} h(x,t)\, e^{-ikx}\, dx \qquad (13.15)$$

We previously used these transforms to formally solve for $h(x,t)$, in the range of $-\infty \le x \le +\infty$, and $t \ge 0$, under the initial condition that $h(x,0) = f(x)$. Recall from Chapter 12 that the wavenumber spectrum on the left hand side of (13.15) has the form

$$F(k,t) = F(k,0)\, \exp\left(-k^2 \left(\frac{T}{S}\right) t\right) \qquad (13.16)$$

where the term for the wave number spectrum, $F(k,0)$, on the right hand side of (13.16) is obtained from the inverse Fourier transform of the initial condition for the hydraulic head $h(x,0) = f(x)$ according to

$$F(k,0) = \int_{-\infty}^{+\infty} f(x)\, e^{-ikx}\, dx \qquad (13.17)$$

In other words, substituting (13.17) for $F(k,0)$ in (13.16), and rearranging terms, leads to

$$F(k,t) = \int_{-\infty}^{+\infty} f(x)\, e^{-ikx} e^{-k^2(T/S)t}\, dx \qquad (13.18)$$

Substituting the right hand side of (13.18) for $F(k,t)$ in (13.14), we obtain

$$h(x,t) = \frac{1}{2\pi} \int_{-\infty}^{+\infty} \left[\int_{-\infty}^{+\infty} f(x')\, e^{-ikx'} e^{-k^2(T/S)t}\, dx' \right] e^{ikx}\, dk \qquad (13.19)$$

which can be rearranged to the form

$$h(x,t) = \frac{1}{2\pi} \int_{-\infty}^{+\infty} \int_{-\infty}^{+\infty} f(x')\, e^{-ik(x-x')} e^{-k^2(T/S)t}\, dx'\, dk \qquad (13.20)$$

Switching the order of integration

$$h(x,t) = \int_{-\infty}^{+\infty} f(x') \left[\frac{1}{2\pi} \int_{-\infty}^{+\infty} e^{-ik(x-x')} e^{-k^2(T/S)t}\, dk \right] dx' \qquad (13.21)$$

we note that the integral in the square bracket is the same integral identity employed in Chapter 12 with a simple change in the variable (x–x'), so that this term becomes

$$\frac{1}{2\pi} \int_{-\infty}^{+\infty} e^{-ik(x-x')} \, e^{-k^2(T/S)t} \, dk \; = \; \frac{1}{\sqrt{\pi}} \sqrt{\frac{S}{4tT}} \, \exp\left(-\frac{(x-x')^2 S}{4tT}\right) \qquad (13.22)$$

However this expression is precisely the impulse response function for a source located at x = x', using the notation

$$\delta_h(x-x',t) \; = \; \frac{1}{\sqrt{\pi}} \sqrt{\frac{S}{4tT}} \, \exp\left(-\frac{(x-x')^2 S}{4tT}\right) \qquad (13.23)$$

(For example, see (13.10) with $x' = x_o$, or (13.13) with $x' = x_i$.)

The convolution integral. Substituting (13.23) for the square bracket in (13.21) leads to

$$h(x,t) = \int_{-\infty}^{+\infty} f(x') \, \delta_h(x-x',t) \, dx' \qquad (13.24)$$

which has the form of an integral operation known in mathematical physics and engineering as a *convolution*; the integral is called a *convolution integral*.

A little reflection will confirm that this operation is precisely the integral analog of the discrete summation described in (13.12). To see this, assume in (13.24) that $f(x) = \Delta h_c = 1$ over the range $-L/2 \le x \le +L/2$. Because $f(x) = 0$ outside this latter range, one can explicitly specify the limits of integration to be from $-L/2$ to $+L/2$. Under these conditions, (13.24) becomes

$$h(x,t) = \int_{-L/2}^{+L/2} \delta_h(x-x',t) \, dx' \qquad (13.25)$$

This could have been written down directly from (13.12), by assuming the interval between the discrete impulses to be constant, $\Delta x_i = \Delta x$, and allowing the position of each unit impulse to be given by $x_i = [(i-1)\Delta x - L/2]$. Then, forcing n, L and Δx to be coupled to each other such that $n = (L/\Delta x) + 1$, upon taking the limit as Δx becomes small (i.e. as Δx becomes the differential dx), we obtain

$$h(x,t) = \lim_{\Delta x \to 0} \sum_{i=1}^{i=n} \left[\delta_h(x,t,x_i) \, \Delta x \right] = \int_{-L/2}^{+L/2} \delta_h(x-x',t) \, dx' \qquad (13.26)$$

We can now construct the "continuous" convolution version of the discrete superposition considered above relative to Figure 13.3. Substituting the impulse response (13.23) into the convolution (13.26), and performing the integration over the specified range of x', the results are presented in Figure 13.4 as a set of profiles of h(x,t) as a function of x for various times t.

The initial perturbation (at t = 0) in hydraulic head is considered to be zero everywhere except between –5 and +5 where it has unit amplitude (shown by the rectangular dashed line in Figure 13.4 labelled h(x,0)).

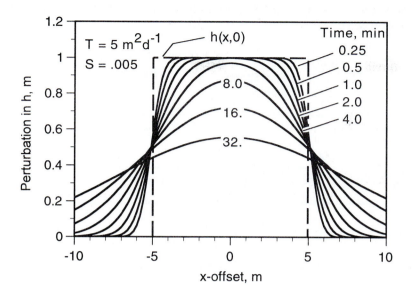

Figure 13.4 The result at various times of a convolution over a continuous distribution of unit impulse response functions spanning the range $-5 \leq x \leq +5$. The rectangular dashed line labelled $h(x,0)$ denotes the initial distribution of hydraulic head at $t = 0$.

This figure shows that for $t > 0$ the sharp edges erode, the amplitude of the perturbation decays over its initial range, and the disturbance diffuses outward into the adjacent medium.

Superposition of Elementary Functions to Emulate Boundary Conditions

An important application of the superposition of elementary solutions is to emulate boundary conditions. Consider, as a simple example, the case where one wants to represent a perturbation in hydraulic head adjacent to a point where the hydraulic head is held at a fixed value; in other words, the amplitude of the perturbation at this point is forced to be zero for all times $t \geq 0$.

We assume, for this example, that the primary unit impulses in h is centered at $x = 1$ m from the edge of a confined aquifer at which point $(x = 0)$ the hydraulic head is explicitly maintained at a constant value of h_0 as shown in the top panel of Figure 13.5. (Constant head is maintained at only the point $x = 0$.)

While this problem can be solved directly, it is much easier to solve it indirectly using a mathematical simulation. The initial values and boundary conditions for the "true" problem in the region $x \geq 0$ can be represented in every way by substituting the case of an *infinite* confined aquifer at an initial static head of h_0 on which we impose not only the *original* unit impulse disturbance at $x = 1$ m, but also *superimpose* an identical, but negative unit impulse simultaneously at $x = -1$ m. Mathematically stated, for the original unit impulse at $x = x_0$ (in this case $x_0 = 1$ m), the composite disturbance is given by

$$h(x,t) = \frac{1}{\sqrt{\pi}} \sqrt{\frac{S}{4tT}} \exp\left(-\frac{(x - x_0)^2 S}{4tT}\right)$$

$$- \frac{1}{\sqrt{\pi}} \sqrt{\frac{S}{4tT}} \exp\left(-\frac{(x + x_0)^2 S}{4tT}\right) \tag{13.27}$$

for $x \geq 0$ and $t \geq 0$, (where as before we suppress the static head term h_0 on the right hand side of (13.27), but understand that its presence is implied).

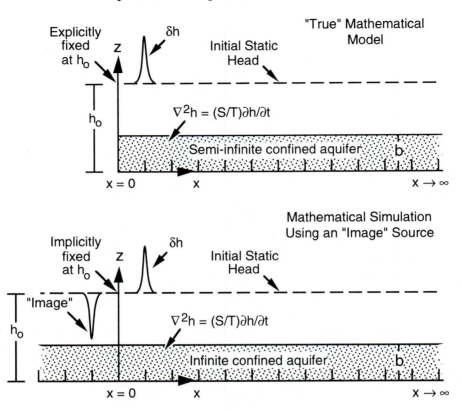

Figure 13.5 *Top panel:* True mathematical situation of a semi-infinite aquifer with a unit impulse source located 1 m in from the edge of the aquifer at $x = 0$. The hydraulic head at $x = 0$ is held constant at h_0 for all $t \geq 0$. *Bottom panel:* Mathematical simulation using an infinite confined aquifer with a simultaneous negative "image" at $x = -1$.

The range for which the solution applies along the x axis is restricted, since the information for $x < 0$ is nonsense in terms of the problem as originally stated (and as illustrated in the top panel of Figure 13.5). The results of such a synthesis are summarized in Figure 13.6, where we illustrate the individual contributions of the two terms in (13.27) as well as their composite. Whereas each of the individual terms are generally non-zero at the origin, their composite is obviously *always* zero at the origin.

This is underscored in Figure 13.7 where the composite response is shown at various times for $t \geq 0$. Note that although, for mathematical interest, the disturbance is shown for both *negative* and *positive* values of x, only the behavior for $x \geq 0$ has physical meaning in the context of the original problem. One can clearly see the diffusion of the disturbance propagating toward increasing values of x, while the hydraulic head is strictly held at 0 (or h_0 for the real case) at all times.

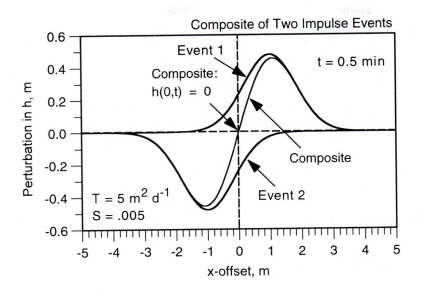

Figure 13.6 Emulating the boundary condition of h_o = constant at the edge of a *semi-infinite* aquifer by using an *infinite* confined aquifer on which we superimpose the *original* unit impulse disturbance at x = 1 m and, simultaneously, an identical, but negative unit impulse at x = −1 m.

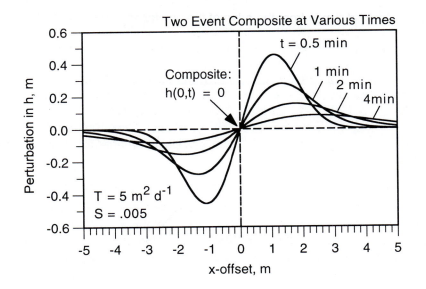

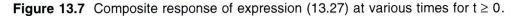

Figure 13.7 Composite response of expression (13.27) at various times for t ≥ 0.

Decay from a Global Initial Condition

Up to this point, this chapter has considered the case of superimposing elementary impulse response functions (also known as "Green's functions") over a relatively local range of the x axis to make up particular composite functions. We now take up the case of superimposing the composite functions themselves to synthesize the response of a system to what we will term "global" initial conditions. We use the term "global" to indicate that an initial condition is being applied over an entire semi-infinite or infinite range of x, rather than at a local point or local region as was done above for unit impulse forcing terms.

Application to a Uniform Offset in the Hydraulic Head

We consider the case in Figure 13.8 where the initial value of the hydraulic head is uniformly displaced over the semi-infinite x axis ($x > 0$) to a value of $h(x,0) = h_1$, but that its value at the origin $x = 0$, is held fixed at h_o for all time $t \geq 0$. In physical terms this might represent the situation where, for example, due to rapid infiltration from precipitation or snow melt, the hydraulic head is suddenly elevated to a new value over a large area adjacent to a reservoir, lake or river, which is held at the static head h_o.

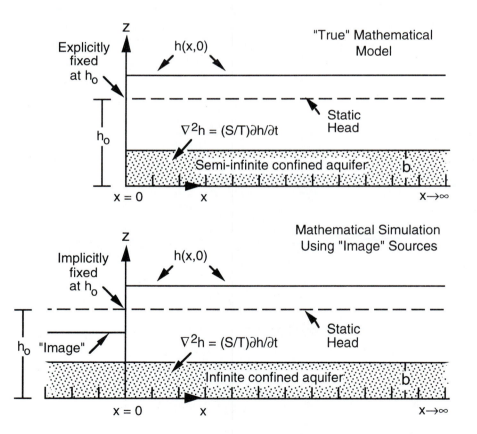

Figure 13.8 *Top panel:* The true mathematical situation of a *semi-infinite* aquifer with an initial hydraulic head offset by a uniform value compared to a pre-transient static level for x > 0. The hydraulic head at x = 0 is held fixed at h_o for all time. ***Bottom panel:*** The mathematical simulation using an *infinite* confined aquifer with an "image" uniform offset for x < 0.

We explore how this function evolves in space and time for $t \geq 0$. To simplify the discussion, assume for now that the perturbation in the hydraulic head is unity, or $h_1 - h_o = \Delta h_c = 1$.

Solution in Terms of the Error Function

Employing the convolution integral in (13.24), we note that a suitable elementary function is the composite impulse response given by (13.27), replacing x_o by the variable of integration x', and limiting the integration to just the positive x axis, such that we obtain the form

$$h(x,t) = \frac{1}{\sqrt{\pi}} \int_0^{+\infty} \left[\sqrt{\frac{S}{4tT}} \exp\left(-\frac{(x-x')^2 S}{4tT}\right) \right.$$

$$\left. - \sqrt{\frac{S}{4tT}} \exp\left(-\frac{(x+x')^2 S}{4tT}\right) \right] dx' \tag{13.28}$$

(Having suppressed the h_o term on the right hand side.) Breaking the integral into two parts according to the two terms of the integrand in (13.28), upon making the following change of variables in the first integral

$$u \sqrt{\frac{4tT}{S}} = x' - x \tag{13.29a}$$

and the following change of variable in the second integral

$$u \sqrt{\frac{4tT}{S}} = x' + x \tag{13.29b}$$

(13.28) can be rewritten in the form

$$h(x,t) = \frac{1}{\sqrt{\pi}} \int_{-\xi}^{+\infty} e^{-u^2} du - \frac{1}{\sqrt{\pi}} \int_{+\xi}^{+\infty} e^{-u^2} du \tag{13.30}$$

where, to simplify notation, we have recalled from Chapter 12 the definition of the parameter

$$\xi = \sqrt{\frac{x^2 S}{4tT}} \tag{13.31}$$

After subtracting the second integral from the first in (13.30), the expression reduces to the form

$$h(x,t) = \frac{1}{\sqrt{\pi}} \int_{-\xi}^{+\xi} e^{-u^2} du \tag{13.32}$$

But considering that the integrand is an even function of u, we can further reduce (13.32) to the form

$$h(x,t) = \frac{2}{\sqrt{\pi}} \int_0^{+\xi} e^{-u^2} du \qquad (13.33)$$

where the right hand side of (13.33) is the well known error function, given by

$$erf(\xi) = \frac{2}{\sqrt{\pi}} \int_0^{+\xi} e^{-u^2} du \qquad (13.34)$$

Recall that, in arriving at the solution (13.33), we assumed the initial perturbation in the hydraulic head was unity, or $h_1 - h_0 = \Delta h_c = 1$. We can restore the statement of the problem to more general values for h_1 and h_0, (and also resurrect the usually suppressed static head term on the right hand side) by expressing (13.33) in the form

$$h(x,t) = h_0 + (h_1 - h_0) \, erf(\xi) \qquad (13.35)$$

In practice, both the amplitude of the transient response and the distance-time parameter ξ are often presented in the following non-dimensional invariant forms to make such analyses more generally useful:

$$\frac{h(x,t) - h_0}{h_1 - h_0} = erf(\xi) \qquad (13.36)$$

Digression on the Properties of the Error Function

The error function $erf(\xi)$ in (13.34) is closely related to the cumulative area under the curve for the normal probability distribution. The constant before the integral sign is simply a normalizing constant so that as ξ approaches infinity, $erf(\xi)$ approaches 1.

Series form. An alternative form for the error function is the series

$$erf(\xi) = \frac{2}{\sqrt{\pi}} \left[\xi - \frac{\xi^3}{3} + \frac{\xi^5}{2! \, 5} - \frac{\xi^7}{3! \, 7} + \cdots \right] \qquad (13.37)$$

A simple substitution of (13.37) into the diffusion equation

$$\frac{\partial^2 h}{\partial x^2} = \frac{S}{T} \frac{\partial h}{\partial t} \qquad (13.38)$$

will show that it is an appropriate solution. (To see this qualitatively, differentiate the first few terms of (13.37) according to (13.38), then neglect higher order terms by assuming that $\xi \ll 1$.)

A plot of the error function erf(ξ) is shown in Figure 13.9.

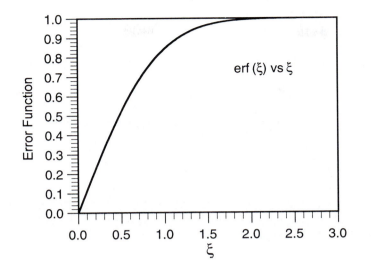

Figure 13.9 The error function erf(ξ) versus ξ.

Thus the plot of erf (ξ) versus ξ in Figure 13.9 is also a plot of the normalized transient head parameter given by (13.36) plotted against the invariant hydraulic parameter.

Asymptotic forms. From the series expansion in (13.37) it is clear that for small ξ (corresponding to small distances or late times), (13.36) reduces to

$$\frac{h(x,t) - h_o}{h_1 - h_o} \approx \xi \tag{13.39}$$

On the other hand, from the integral expression in (13.34) or the plot in Figure 13.9, it is clear that for large ξ (corresponding to large distances, or early times), (13.36) reduces to

$$\frac{h(x,t) - h_o}{h_1 - h_o} \approx 1 \tag{13.40}$$

The latter implies that, at these values of x and t, the recovery to the static level h_o is not yet significant, and the initial conditions still dominate.

Application to a Step Offset in the Hydraulic Head at x = 0

The error function as a general solution. While we arrived at the error function as a solution to the specific initial value problem discussed above, this form can, in fact, be used to represent solutions to other initial and boundary value conditions. In other words, for these other problems, the fundamental solution might be in terms of the *error function*, but the statement of the initial conditions and forcing term is different. In such a case, the error function can be used to represent how the *dynamic* component of the disturbed hydraulic head evolves in space, $x \geq 0$, and time, $t \geq 0$, providing an appropriate adjustment is made to the constant terms and coefficient to meet the assumed initial and boundary conditions.

Statement of the step offset problem. To illustrate this approach, consider the case in Figure 13.10 where the initial value of the hydraulic head over the semi-infinite x axis (x > 0) is its static value h_0. At t = 0 the hydraulic head at the origin, x = 0, suddenly drops to a new value of $h(0,t) = h_1$, and is held thus for all time t ≥ 0.

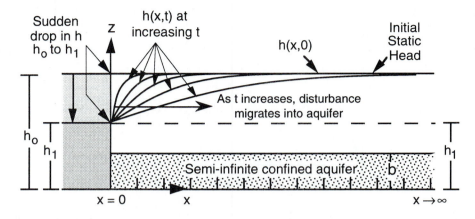

Figure 13.10 Illustrating a step offset from h_0 to h_1 in the hydraulic head at x = 0, starting at t = 0 and persisting for all t ≥ 0.

As time progresses, and the head at x = 0 is held fixed at its new value h_1, one would expect the disturbance to propagate into the aquifer as shown in Figure 13.10. It should be apparent that the shape of the perturbation is similar to the plot of the error function in Figure 13.9. Is the error function actually a solution to this problem?

Noting that expressions of the form (13.35) are solutions to the 1-D transient flow equation, we write a generic solution to the latter equation in the form

$$h(x,t) = h_c + \Delta h_c \, erf(\xi) \tag{13.41}$$

where h_c and Δh_c are constant in x and t, and are determined by the appropriate initial and boundary value conditions.

The particular solution of the 1-D transient flow equation that meets the initial and boundary value conditions of the step offset problem specified above, and illustrated in Figure 13.10, is obtained if in (13.41) we let $h_c = h_1$, and $\Delta h_c = -(h_1 - h_0)$, so that

$$h(x,t) = h_1 - (h_1 - h_0) \, erf(\xi) \tag{13.42}$$

The general behavior of the latter expression can be appreciated from inspecting the plot of the error function in Figure 13.9, and noting that (13.42) reduces to

$$h(x,t) \approx h_1 \quad \text{for small x; } t = \text{constant} \tag{13.43a}$$

$$h(x,t) \approx h_0 \quad \text{for large x; } t = \text{constant} \tag{13.43b}$$

$$h(x,t) \approx h_0 \quad \text{for small t; } x = \text{constant} \tag{13.43c}$$

$$h(x,t) \approx h_1 \quad \text{for large t; } x = \text{constant} \tag{13.43d}$$

Rearranging (13.42) in the form

$$\frac{h_1 - h(x,t)}{h_1 - h_o} = \text{erf}(\xi) \tag{13.44}$$

one can clearly see that Figure 13.9 might be used to represent the erosion of the sharp initial edge between h_1 and h_o for a profile of h into the aquifer at a fixed value of time. For $\xi \leq .05$, the transient will have decayed to within approximately 5% of its *final* steady-state value, h_1. For $\xi > 0.6$, erf (ξ) rapidly approaches its asymptotic value of unity. As examples

$$\xi = 1.0, \text{erf}(\xi) = 0.84;$$
$$\xi = 1.2, \text{erf}(\xi) = 0.91;$$
$$\xi = 1.4, \text{erf}(\xi) = 0.95;$$
$$\xi = 2.0, \text{erf}(\xi) = 0.995.$$

This in turn implies that for values of ξ on the order of 1 and larger, erf (ξ) is approximately unity and, from (13.44), we have

$$\frac{h_1 - h(x,t)}{h_1 - h_o} \approx 1 \tag{13.45}$$

implying that

$$h(x,t) \approx h_o \tag{13.46}$$

which is to say that the transient condition has not yet reached those values of x and t corresponding to $\xi \geq 1$; x is too large and/or t is too small for $h(x,t)$ to have changed sensibly from its pre-transient static value of h_o.

Complementary Error Function

Relation to error function. In representing the transient response of an aquifer to the step offset at $x = 0$ discussed in the last section, it is sometimes useful to couch the perturbation of the hydraulic head in terms relative to its initial steady-state condition h_o, rather than in terms relative to h_1 as in the last section. Accordingly, we now derive a dynamic expression for

$$\delta h = h(x,t) - h_o \tag{13.47}$$

which represents the difference between the hydraulic head $h(x,t)$ and its initial steady-state condition h_o. This is done by employing the following complementary form.

Rewrite (13.42) in the form

$$h(x,t) - h_1 = - (h_1 - h_o) \, \text{erf}(\xi) \tag{13.48}$$

Add $(h_1 - h_o)$ to both sides of (13.48), and factor, to obtain the following invariant form

$$\frac{h(x,t) - h_o}{h_1 - h_o} = 1 - \text{erf}(\xi) \qquad (13.49)$$

The term on the right side of (13.49) appears so frequently in the literature that it is given a special name, the "complementary error function" *erfc(ξ)*, defined as

$$\text{erfc}(\xi) = 1 - \text{erf}(\xi) \qquad (13.50)$$

Thus (13.49) can be written in the form

$$\frac{h(x,t) - h_o}{h_1 - h_o} = \text{erfc}(\xi) \qquad (13.51)$$

where, as before, ξ is given by (13.31). The behavior of the complementary error function $\text{erfc}(\xi)$ as a function of ξ is shown in Figure 13.11.

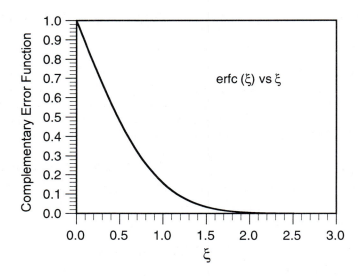

Figure 13.11 The complementary error function erfc (ξ) versus ξ.

From (13.50) and the definition of the error function from expression (13.34), it should be clear that the complementary error function erfc (ξ) is related to the error integral by

$$\text{erfc}(\xi) = \frac{2}{\sqrt{\pi}} \int_{\xi}^{\infty} e^{-u^2} du \qquad (13.52)$$

and has the following attributes:

- erfc (ξ) approaches 0 as ξ approaches ∞. (There is not much area under this segment of the $\exp(-u^2)$ curve).

- erfc (ξ) approaches 1 as ξ approaches 0. (All the area is under this segment of the $\exp(-u^2)$ curve).

Asymptotic Behavior of h

Expression (13.51) has the following properties:

1) For early times (or large x), erfc (ξ) approaches 0, and

$$h(x,t) \approx h_0 \qquad (13.53)$$

where h_0 is the initial pre-transient static head.

2) For late times (or small x), erfc (ξ) approaches 1, and

$$h(x,t) \approx h_1 \qquad (13.54)$$

where h_1 is the final, or imposed, head condition.

Alternative Representations

For some applications, there are more convenient formats for plotting such transients.

Alternative A. If one is investigating the behavior of h versus time (at a fixed distance x), the disadvantage of plotting an expression like (13.51) versus ξ — the usual convention in mathematics and physics — is that it is somewhat confusing to have a function changing from left to right while time is implicitly *increasing* from right to left. One usually thinks in terms of time evolving to the right, so it may be disconcerting to many to have the *new* steady-state condition in Figure 13.11 asymptotic to the *left hand side* of the plots.

This concern was addressed by Theis (1935) who plotted the relevant hydraulic head parameters as functions of $(\xi)^{-1}$, rather than ξ as is usually done in mathematical physics. Figure 13.12 is such a plot.

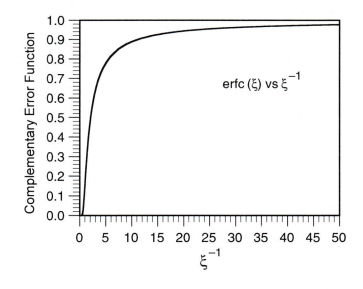

Figure 13.12 Complementary error function erfc (ξ) versus $1/\xi$.

This figure illustrates that by plotting the transient hydraulic head against

$$(\xi)^{-1} = \sqrt{\frac{4tT}{x^2 S}} \tag{13.55}$$

along the abscissa, one can more readily visualize the temporal evolution of the phenomena as the square root of time increases to the right.

Moreover, Figure 13.12 shows that when one is at a fixed distance from the source, the invariant parameter $1/\xi$ must approach very large values (here shown as 50 or greater), for the value of the perturbed hydraulic head to asymptotically approach its final steady-state value ($h \approx h_1$). However, if one is satisfied with only approaching 90% of the final value, then the invariant parameter need be only $1/\xi \approx 10$.

Alternative B. It is sometimes more useful to plot these results using logarithmic coordinates. Such a plot, using the same data as above, is shown in Figure 13.13.

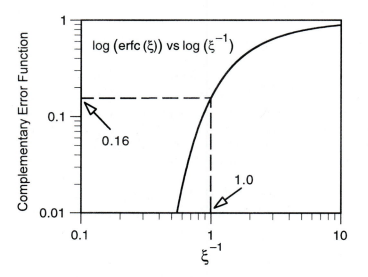

Figure 13.13 Log-log plot of the complementary error function erfc (ξ) versus $1/\xi$. The dashed lines and arrow show the amplitude in log-log space of the complementary error function erfc (ξ) after one "time constant" $1/\xi = 1.0$.

This figure clearly shows that most of the transient effect (from approximately 1% to 90% of the final value) occurs within the range of

$$0.55 < (\xi)^{-1} < 10 \tag{13.56}$$

It is also clear from Figure 13.13 that the recovery rate at late time is very slow — recall that the abscissa is increasing as $t^{1/2}$. One definition of a characteristic time constant for the recovery of the transient perturbation is based on setting the invariant parameter ξ^{-1} to unity, or

$$(\xi)^{-1} = 1.0 \tag{13.57}$$

from which, for a given x, S and T, we solve for

$$t_c = \frac{x^2 S}{4 T} \tag{13.58}$$

As shown in Figure 13.13, one can see that in one time constant t_c, the transient has changed approximately 16% from its pre-transient static condition.

At late time, however, we see that the "rate of recovery" is much reduced. For example, to achieve 90% of its final steady-state value takes somewhat longer than 100 time constants! Thus, while Figures 13.12 and 13.13 show that conditions change relatively rapidly in early time, full recovery to a final steady-state value at late time can be agonizingly slow.

Hydraulic Diffusivity

Occasionally, authors couch ξ in terms of the *hydraulic diffusivity*, defined by

$$\kappa = T/S \tag{13.59}$$

so that

$$\xi = \sqrt{\frac{x^2}{4 \kappa t}} \tag{13.60}$$

COMPARING PERIODIC AND APERIODIC TRANSIENT FLOW EVENTS

Scaling Parameters in Periodic and Aperiodic Flow

Aperiodic flow is governed by the non-dimensional scale length parameter

$$\xi = \sqrt{\frac{x^2 S}{4 t T}} \tag{13.31}$$

An analogous parameter can be determined for *periodic* flow, beginning with the attenuation length x_c, given in the last chapter by

$$x_c = \sqrt{\frac{2T}{\omega S}} \tag{13.61}$$

We earlier defined a characteristic period for the fluctuation of $T_o = 2\pi/\omega$, where ω is the radian frequency of the driving source. One can calculate a non-dimensional scale length ξ_p for periodic flow by normalizing the true distance x by the characteristic attenuation length x_c. This leads to

$$\xi_p = \left(\frac{x}{x_c}\right) = \sqrt{\frac{\pi x^2 S}{T_o T}} \tag{13.62}$$

The similarity in form between ξ for aperiodic flow given by (13.31) and ξ_p for periodic flow given by (13.62) is noteworthy. That ξ is a factor of $(4\pi)^{1/2}$ smaller than ξ_p is due to the fact that the aperiodic event, being a step function in time, has a rich component of higher frequencies that, according to (13.61), attenuate more rapidly with x, than the single frequency periodic event.

Example: Aperiodic Transient Flow

The transient aperiodic theory of the last section is applied to the same confined aquifer considered previously in Chapter 12 for the periodic case. The top panel of Figure 13.14 shows the *aperiodic* forcing term as a pulse of constant unit amplitude, starting at time t = 0 and extending to t = 0.5 days (12 hr), at which time the forcing term returns to zero (or its initial static head value of h_o).

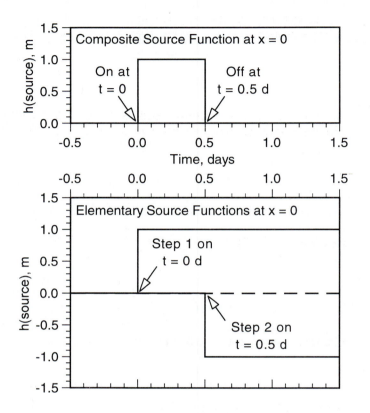

Figure 13.14 *Top panel:* The aperiodic forcing term used to "drive" transient flow in a confined aquifer discussed in the text. *Bottom panel:* The two elementary step functions that are superimposed to simulate the composite pulse in the top panel.

The composite pulse is simulated computationally with the two step functions shown in the bottom panel of Figure 13.14. A positive step of unit amplitude for t ≥ 0 hr represents the disturbance being "turned on". The disturbance being "turned off" is simulated by adding a second equal but opposite (negative) event at the origin (x = 0) for t ≥ 12 hr. While both steps continue on to infinite time at their respective levels, their sum total causes the composite disturbance to be reset to zero at x = 0, for all t ≥ 12 hr. Such a synthesis is a very common procedure in mathematical physics.

Practical examples of such a phenomenon might be a flood event on land, or a storm surge at the coast, where one would want to study the propagation of such a pulse into an aquifer, and how its form changes in space and time. The characteristic time of 0.5 d (12 hr) for the aperiodic impulse was chosen simply to have a common reference to the 12 hour *periodic* source considered in Chapter 12. Figure 13.15 illustrates the propagation of the perturbation into the confined aquifer at various times after t = 0, when the pulse is "turned on". Clearly there is a diffusion-like migration of the disturbed h into the aquifer at various time slices. At t = 0.5 d (12 hr), the composite hydraulic head is reset to the static value h_o at x = 0, for all t ≥ 12 hr, but continues to propagate into the aquifer as a diffusive signal as shown in the bottom panels of Figure 13.15.

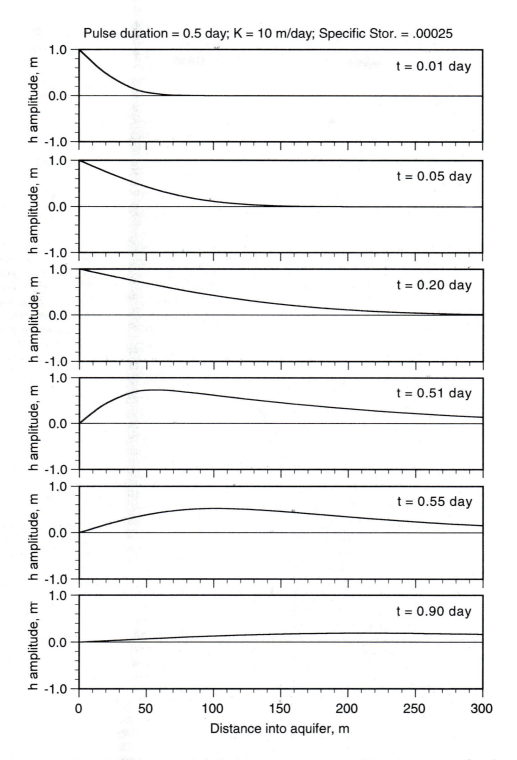

Figure 13.15 The propagation of a disturbance in the hydraulic head into a confined aquifer due to the aperiodic forcing term illustrated in Figure 13.14. The hydraulic parameters are the same as the periodic case illustrated in Figure 12.2 ($K = 10$ m day^{-1} and $S_s = 0.00025$).

Chapter 14. Transient Well Discharge from a Confined Aquifer

TWO DIMENSIONAL FLOW TO A WELL

Recalling Poisson's Equation for Radial Flow

In Chapter 10, we considered some fundamental aspects of two-dimensional horizontal flow to a well discharging from a confined aquifer. With reference to Figure 10.1, we developed an expression for radial flow with sources of water in a confined aquifer in the form of a Poisson's equation

$$\frac{d}{dr}\left(\frac{dh}{dr}\right) + \frac{1}{r}\frac{dh}{dr} = -\frac{wb}{T} \qquad (10.34)$$

The term on the left follows from substituting Darcy's law into the divergence relation for the depth-integrated Q_r, and factoring out the transmissivity T. The right hand side of the equation represents all the sources (or sinks) of water through the entire thickness of the aquifer, divided by its transmissivity T. Recall that w is a generic water source term that represents the volume of water produced per unit volume of the aquifer [having dimensions of $L^3 L^{-3} T^{-1}$, or simply T^{-1}]. To obtain the water productivity over the thickness of the aquifer, we multiply w by the thickness of the aquifer b. While in Chapter 10 we considered this as a steady-state source of fluid, according to (11.3) this is precisely the definition of water released through a transient change in storage. In other words, from (11.46) the fractional volume of water expelled from a volume of the aquifer due to a change in storage is given by

$$w = (1/V_t)\partial V_w/\partial t = -\partial S_w/\partial t \qquad (14.1)$$

Diffusion Equation for Radial Flow

We showed in Chapter 11 that the change in fractional volume of water in storage is related to a change in hydraulic head through

$$dV_w/V_t = -S_s\, dh \qquad (11.44)$$

where S_s is the specific storage given by

$$S_s = [\alpha + n\beta]\, \rho g. \qquad (11.43)$$

and where α is the compressibility of the matrix material, β is the compressibility of water, n is the porosity, ρ is the density and g is the force of gravity per unit mass. Thus (14.1), along with (11.44) leads to the following form for the source term in (10.34)

$$w = -S_s\partial h/\partial t \qquad (14.2)$$

The storativity S is the depth-integrated specific storage given for a uniform confined aquifer of thickness b given by

$$S = b\, S_s \qquad (14.3)$$

Thus the flow equation represented by (10.34) can be modified to the following expression

$$\frac{\partial^2 h}{\partial r^2} + \frac{1}{r}\frac{\partial h}{\partial r} = \frac{S}{T}\frac{\partial h}{\partial t} \qquad (14.4)$$

to account for loss (or gain) of water in the aquifer due to elastic deformation associated with a transient variation in the hydraulic head.

Relation (14.4) is our fundamental relation for the flow in a confined aquifer induced by the discharge (or recharge) from a fully penetrating well at the origin.

Radial Flow Close to a Discharging Well

The pump discharges fluid at the origin at the rate Q. Conservation of mass requires that the total discharge Q at $r = 0$ results in a specific discharge q_r at the walls of the well of

$$\lim_{r \to 0}(r\, q_r) = -\frac{Q}{2\pi b} \qquad (14.5)$$

where b is the thickness of the aquifer. Since, from Darcy's law

$$q_r = -K\frac{\partial h}{\partial r} \qquad (14.6)$$

Relation (14.5) can be expressed in terms of the gradient of the hydraulic head by

$$\lim_{r \to 0}\left(r\frac{\partial h}{\partial r}\right) = \frac{Q}{2\pi T} \qquad (14.7)$$

where T is the transmissivity given by $T = bK$. Our agenda in the following discussion is to match solutions to (14.4) to a condition on the derivative of h at the origin represented by (14.7), as well as to other conditions on h specified below. We first need solutions to the differential form (14.4).

SOLUTION TO THE RADIAL FLOW EQUATION

Separation of Variables

We seek solutions to (14.4) of the form

$$h(r,t) = R(r)\,T(t) \qquad (14.8)$$

where $R(r)$ and $T(t)$ are functions exclusively of r and t, respectively. As in Chapter 12, the use of $T(t)$ to represent the function of time follows standard mathematical practice, but the reader needs to exercise some caution so as not confuse this notation with T used for the transmissivity. The time function will always be associated with an explicit time dependence shown by brackets.

Following an analogous procedure to that used earlier in Chapter 12, substituting (14.8) into (14.4) leads to

$$\frac{1}{R(r)} \left[\frac{\partial^2 R(r)}{\partial r^2} + \frac{1}{r} \frac{\partial R(r)}{\partial r} \right] = \frac{S}{T} \frac{1}{T(t)} \frac{\partial T(t)}{\partial t} = -k^2 \qquad (14.9)$$

where $-k^2$ is the separation constant (Margenau and Murphy, 1956).

The first order differential equation in time

$$\frac{dT(t)}{dt} = -k^2 \frac{T}{S} T(t) \qquad (14.10)$$

has a solution of the form

$$T(t) = const \cdot e^{-\alpha t} \qquad (14.11)$$

where α is a damping constant, such that

$$\alpha = k^2 T / S \qquad (14.12)$$

The second-order differential equation in r becomes

$$r^2 \frac{d^2 R(r)}{dr^2} + r \frac{dR(r)}{dr} + k^2 r^2 R(r) = 0 \qquad (14.13)$$

which is Bessel's equation (Abramowitz and Stegun, 1965), having the solution

$$R(r) = J_0(kr) \qquad (14.14)$$

where $J_0(kr)$ is the zero-th order Bessel function of the first kind, plotted below.

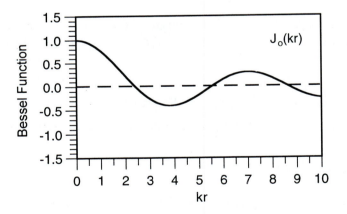

Figure 14.1 A zero order Bessel function of the first kind, $J_0(kr)$, as a function of kr.

Therefore, the general solution to our original diffusion equation (14.4) is given by the superposition of the product $h(r,t) = R(r) T(t)$ over all k, where $R(r) = J_0(kr)$, and $T(t) = const \cdot \exp(-\alpha t)$, respectively, and where $\alpha = k^2 (T/S)$.

Thus $h(r,t)$ becomes

$$h(r,t) = const \cdot J_o(kr) \exp[-(k^2T/S)t] \qquad (14.15)$$

Since *any* k is an allowed solution at this point, *all* k's might be solutions, which through linear superposition might be represented by

$$h(r,t) = const_1 \cdot J_0(k_1r) \exp[-(k_1^2T/S)t]$$
$$+ const_2 \cdot J_0(k_2r) \exp[-(k_2^2T/S)t]$$
$$+ const_3 \cdot J_0(k_3r) \exp[-(k_3^2T/S)t] + \cdots \qquad (14.16)$$

and so forth, where the constants, $const_j$, and the wavenumbers, k_j, depend on the specifics of the problem, namely the initial and boundary conditions.

While the latter expression may be appropriate for a few special problems, in general rather than a "summation" over discrete values of k, the solution is better expressed as an integration over the continuous variable k, thus obtaining

$$h(r,t) = \int_0^\infty C(k) \, J_o(kr) \exp[-(k^2T/S)t] \, dk \qquad (14.17)$$

For future convenience, letting $C(k) = k \, G(k)$, (14.17) can be written

$$h(r,t) = \int_0^\infty G(k) \, J_o(kr) \exp[-(k^2T/S)t] \, k \, dk \qquad (14.18)$$

which is our desired solution to the flow equation (14.4).

We need now only apply the initial and boundary conditions relevant to our particular problem to determine the kernel function G(k), and complete the integration over k.

Initial and Boundary Conditions

The initial and boundary conditions relevant to a well discharging at the origin starting at $t = 0$ are the following:

- Initially, $h(r,0) = h_o$, for all r, $0 < r < \infty$ $\qquad (14.19)$

- For $t > 0$, pumping proceeds at a constant discharge Q. As shown above for (14.7), conservation of mass and Darcy's law leads to the following condition on the gradient of the hydraulic head

$$\lim_{r \to 0} \left(r \frac{\partial h}{\partial r} \right) = \frac{Q}{2\pi T} \qquad (14.20)$$

where T is the transmissivity given by $T = bK$.

- During the pumping operation, a constant hydraulic head is maintained at infinite r, or

$$h(\infty, t) = h_o, \text{ for all t.} \qquad (14.21)$$

Conditions at the Origin and Early Time

We first consider the conditions at $t \approx 0$ close to the origin $r = 0$. For $t \approx 0$,

$$\exp[-(k^2 T/S)t] \approx 1 \qquad (14.22)$$

and our above solution (14.18) reduces to

$$h(r) = \int_0^\infty G(k)\, J_0(kr)\, k\, dk \qquad (14.23)$$

Expression (14.23) can be looked upon as one member of a Fourier-Bessel integral transform pair (Margenau and Murphy, 1956), where the second member — the kernel function $G(k)$ — is given by

$$G(k) = \int_0^\infty h(r)\, J_0(kr)\, r\, dr \qquad (14.24)$$

Note: Relations (14.23) and (14.24) are also known as a "Hankel integral transform pair" (Lebedev, 1972; p.130), and are completely analogous to the Fourier integral transform pair discussed in Chapter 12.

Impulse Source Function

To determine $G(k)$ from (14.24), we next consider (still at $t \approx 0$) the effects of an impulse of water being withdrawn at a constant discharge rate Q over a small time increment Δt from the aquifer at the origin. To be specific, the total volume drawn will be given by $Q\Delta t$ (in units of L^3).

From our discussion at the beginning of the chapter, we assume that the water is withdrawn from elastic storage in the aquifer, and is compensated for by a reduction in hydraulic head according to

$$\frac{1}{V_t}\frac{\partial V_w}{\partial t} = -S_s\frac{\partial h}{\partial t} \qquad (14.25)$$

The infinitesimal volume fraction of water removed from a reference volume of the aquifer over a differential time increment dt is obtained from (14.25) as

$$\frac{1}{V_t}\frac{\partial V_w}{\partial t}\, dt = -S_s\frac{\partial h}{\partial t}\, dt \qquad (14.26)$$

Assuming the pump discharges water from the well at a constant rate of Q, beginning instantaneously at $t = 0$, and lasting an infinitesimal period of time Δt, implies that the integral of (14.26) over space (the entire volume of the aquifer), and over time (from $t = 0$ to Δt) is equal to the total volume of the impulse of water discharged.

Mathematically, this places the following condition on such an integration:

$$\int_0^{\Delta t} \left[\iiint_{\substack{\text{Vol of} \\ \text{aquifer}}} \frac{1}{V_t} \frac{\partial V_w}{\partial t} \, dV \right] dt = - \int_0^{\Delta t} \left[\iiint_{\substack{\text{Vol of} \\ \text{aquifer}}} S_s \frac{\partial h}{\partial t} \, dV \right] dt$$

$$= \int_0^{\Delta t} Q \, dt = Q \Delta t \qquad (14.27)$$

where, as noted above, the last term in (14.27) represents the total volume of fluid discharged over the time increment Δt, which is $Q \Delta t$.

We now explicitly integrate the second term in (14.27) over the entire volume of the aquifer, still assuming that we are dealing with an *impulse* of discharging water at the origin.

Integrating the second term in (14.27) over the thickness of a confined aquifer is equivalent to a multiplication of the integrand by b. We thus introduce the storativity, $S = bS_s$, to represent the integration over the vertical dimension of the aquifer. Integrating over the rest of the volume is done with respect to azimuth θ and radius r, and the expression reduces to

$$\int_{t=0}^{\Delta t} \int_{r=0}^{\infty} \int_{\theta=0}^{2\pi} S \frac{\partial h}{\partial t} r \, dr \, d\theta \, dt = -Q \Delta t \qquad (14.28)$$

where $Q \Delta t$ is the total volume of water discharged from the well.

Let us review conceptually where we are in our discussion. We are attempting to characterize the behavior of flow in the vicinity of the well in the very first instant of time after discharge commences. We would expect, intuitively, or at least in the light of our findings for the 1-D case in Chapter 12, that the initial perturbation in head is restricted to the vicinity of the well — which is to say the *"cone of depression"* is very tightly focussed at the origin. Moreover, as we extract that first increment of water at a constant rate from an infinitesimal radius of influence Δr_{eff}, according to our conservation relations we would expect the drawdown in hydraulic head to be approximately linear in time — and we can make Δt as short as need be to assure this. At each position over the aquifer, we define a perturbation in hydraulic head over the time increment Δt by

$$\delta h(r,\theta) = \int_0^{\Delta t} \frac{\partial h}{\partial t} \, dt \qquad (14.29)$$

The left hand term in expression (14.28) can thus be integrated over time to provide the form

$$\int_{r=0}^{\infty} \int_{\theta=0}^{2\pi} S\, \delta h(r,\theta)\, r\, dr\, d\theta = -Q\,\Delta t \qquad (14.30)$$

We next assume that $\delta h(r,\theta)$ is symmetrical with θ. Integrating over θ, therefore, leads to

$$2\pi \int_{r=0}^{\infty} S\, \delta h(r)\, r\, dr = -Q\,\Delta t \qquad (14.31)$$

which can be rearranged to the form

$$\int_{r=0}^{\infty} \delta h(r)\, r\, dr = -\frac{Q\,\Delta t}{2\pi S} \qquad (14.32)$$

Now assuming that the "impulse function" $\delta h(r)$ is concentrated close to the origin, we substitute $\delta h(r) = \delta_h(r)$, where the latter represents a distribution function concentrated at the origin — analogous to the Dirac delta function discussed in Chapter 12. (14.32) thus becomes

$$\int_{r=0}^{\infty} \delta_h(r)\, r\, dr = -\frac{Q\,\Delta t}{2\pi S} \qquad (14.33)$$

Let us reflect on this last relation. The right hand side represents the total well discharge over a time increment Δt, divided by 2π times the storativity of the aquifer. On the left hand side of the relation, we are assuming that there exists an impulse function $\delta_h(r)$ concentrated at the origin that, when multiplied by r, and integrated over all r, equals the right hand side. At this point, we really do not care about the details of the functional form of $\delta_h(r)$, only that it is concentrated (i.e. non-zero) at the origin and integrates to the right side of (14.33).

Determining the Hankel Transform

We may now use the latter expression (14.33) to evaluate the Hankel transform given above as

$$G(k) = \int_0^{\infty} h(r)\, J_0(kr)\, r\, dr \qquad (14.24)$$

To do so, we substitute $\delta_h(r)$ for $h(r)$, so that the latter expression becomes

$$G(k) = \int_0^\infty \delta_h(r) \, J_o(kr) \, r \, dr \qquad\qquad (14.34)$$

However, because $\delta_h(r)$ is so concentrated at $r \approx 0$, we can assume that over the range of r for which $\delta_h(r)$ is non-zero, $J_o(kr) \approx 1$ (see Figure 14.1). If this is not true for a particular functional form of $\delta_h(r)$ that we have selected, we can make the function more narrow in time, or more focussed at the origin, or select another form for the function. Thus (14.34) becomes

$$G(k) = \int_0^\infty \delta_h(r) \, r \, dr \qquad\qquad (14.35)$$

where upon inspecting (14.33) above, we see that the right hand side of (14.35) can be replaced by the right hand side of (14.33) so that

$$G(k) = -\frac{Q\Delta t}{2\pi S} \qquad\qquad (14.36)$$

Impulse Response in Integral Form

Substituting the right hand side of (14.36) for G(k) in the Hankel transform of (14.23) leads to an *impulse response* for h(r,t) that we denote by

$$\delta_h(r,t) = -\int_0^\infty \frac{Q\Delta t}{2\pi S} \, J_o(kr) \, \exp[-(k^2 T/S)t] \, k \, dk \qquad\qquad (14.37)$$

or

$$\delta_h(r,t) = -\frac{Q\Delta t}{2\pi S} \int_0^\infty J_o(kr) \, \exp[-(k^2 T/S)t] \, k \, dk \qquad\qquad (14.38)$$

An Integral Identity

One can look up the integral on the right of (14.38) in any good table of integrals. It is a special form of Weber's integral (Lebedev, 1972)

$$\int_0^\infty e^{-a^2 k^2} \, J_v(bk) \, k^{v+1} \, dk \qquad\qquad (14.39)$$

which Lebedev (p.133) shows (for the special case $v = 0$) leads to the integral identity

$$\int_0^\infty e^{-a^2 k^2} \, J_o(bk) \, k \, dk = \frac{1}{2a^2} \, e^{-b^2/4a^2} \qquad\qquad (14.40)$$

Impulse Response from Transient Discharge

By inspecting (14.38) and the integral identity (14.40). we now make the following substitutions for the constant parameters in (14.40): a^2 becomes Tt/S; b^2 (not to be confused with the thickness of the aquifer) becomes r^2. Thus, substituting (14.40) for the integral on the right hand side of (14.38) leads to

$$\delta_h(r,t) = -\frac{Q\Delta t}{2\pi S}\frac{S}{2Tt}\exp[-(r^2S)/(4tT)] \qquad (14.41)$$

which upon simplifying becomes

$$\delta_h(r,t) = -\frac{Q\Delta t}{4\pi tT}\exp[-(r^2S)/(4tT)] \qquad (14.42)$$

The latter expression is the behavior of the hydraulic head for an impulsive withdrawal of a volume of fluid $Q\Delta t$ from the origin $r = 0$, at $t = 0$. Whereas we were not very particular about the detailed shape of $\delta_h(r)$ when we were setting up the relation (14.33), we now see that, in fact, it has a very special form driven by the physics of the discharge process. In this sense, $\delta_h(r,t)$ is a very special type of distribution function — it is the 2-D planar "point source" analog of the 1-D impulse response discussed in Chapter 12.

Behavior of Impulse Response in Space and Time

A plot of the impulse response $\delta h(r,t)$ represented by (14.42) is given in Figure 14.2 for various times. The parameters of the aquifer are the same as those used for the example in Chapter 10 with reference to Figure 10.2 and Figure 10.3.

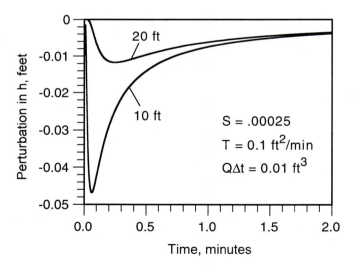

Figure 14.2. The impulse response $\delta h(r,t)$ represented by (14.42) given for various times and distances. The parameters of the aquifer are $T = 0.1$ ft^2/min and $S = 0.0025$; the same as those used for the example in Chapter 10. Q is constant at 1.0 ft^3/min for a $\Delta t = 0.01$ min.

We assume a confined aquifer having a transmissivity of $T = 0.1$ ft^2 min^{-1} (13.4 m^2 day^{-1}) and a storativity of $S = 0.00025$. (As before, we employ English units in deference to common practice among community and regulatory agencies.) For the present purpose, we assume that the impulse in Q is such that the flow rate is held constant at 1.0 ft^3 min^{-1} (0.472 x 10^{-3} m^3 s^{-1}, 0.472 liter s^{-1}, or 40.8 m^3 day^{-1}) for a time increment of $\Delta t = 0.01$ min (6.94 x 10^{-6} day). Referring to Figure 14.2, note the diffusive-like behavior of the impulse as it migrates from r = 10 ft (3.05 m) to r = 20 ft (6.1 m).

TRANSIENT RESPONSE TO CONTINUOUS DISCHARGE

Solution for Constant Discharge $0 < \tau < t$

To determine the hydraulic head for a *continuous* discharge rate of Q [L^3 T^{-1}], we allow Δt to become the differential time increment dt, and convolve over the continuum of impulses from the commencement of discharge to the time of observation. This is to say we integrate (14.42) using a convolution integral (analogous to the procedure in Chapter 13 for the 1-D case) from $\tau = 0$ to $\tau = t$; where t is the particular time sample at which the transient head is being evaluated. Following the discussion in Chapter 13, the resulting composite perturbation is given by

$$\delta h(r,t) = \int_0^t \delta_h(r,\tau)\, d\tau = -\frac{Q}{4\pi T} \int_0^t \frac{e^{-r^2 S/(4\tau T)}}{\tau}\, d\tau \qquad (14.43)$$

which is the change in hydraulic head relative to its initial static value (usually h_o).

The Exponential Integral

We introduce the invariant parameter
$$u = r^2 S /(4 t T) \qquad (14.44)$$

which is known among hydrologists as the Theis (1935) parameter. With the change of variable, $u = r^2 S / 4\tau T$, (14.43) can be rearranged to the form

$$\delta h(r,t) = -\frac{Q}{4\pi T} \int_u^\infty \frac{e^{-u'}}{u'}\, du' \qquad (14.45)$$

One should note that the invariant parameter for radial flow, $u = r^2 S /(4tT)$, is analogous to the square of the invariant parameter for 1-D flow, $\xi^2 = x^2 S /(4tT)$, employed in Chapters 12 and 13.

The definite integral in (14.45) is the well-documented exponential integral $E_1(u)$ (Abramowitz and Stegun, p.228), so that the expression becomes

$$\delta h(r,t) = - (Q/4\pi T)\, E_1(u) \qquad (14.46)$$

Note: $E_1(u)$ is known as the "well function" among hydrologists (Strack, 1989), but is a function commonly used in a wide range of scientific and engineering disciplines studying diffusion-like processes.

In series form, the exponential integral is given by

$$E_1(u) = -0.5772156649... - \ln u - \sum_{n=1}^{\infty} \frac{(-1)^n u^n}{n n!} \qquad (14.47)$$

where the leading constant on the right hand side is the negative of the Euler constant, and the term $\ln u$ is the logarithm of u to the base e. A plot of $E_1(u)$ versus u is shown in the left panel of Figure 14.3, and has the interesting property that, except for the multiplicative factor $(Q/4\pi T)$, this single curve (as did the error integral for the 1-D case in Chapter 13) represents the drawdown properties of the aquifer for all combinations of T, S, distance and time.

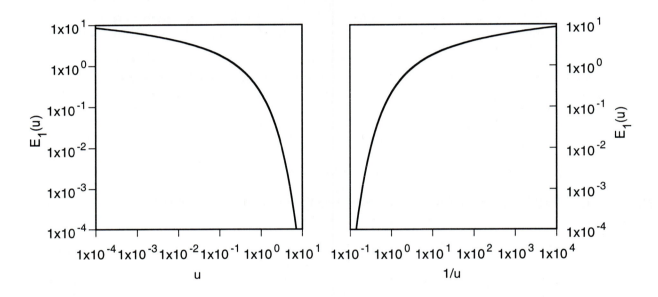

Figure 14.3 *Left panel:* Plot of $E_1(u)$ versus u; the radius-squared increases to the right at fixed time. *Right panel:* Plot of $E_1(u)$ versus 1/u; the time increases to the right at fixed radius.

For fixed time, $E_1(u)$ is plotted versus u in the top panel of Figure 14.3 such that the radius-squared increases to the right. For fixed radius, the time increases to the left.

Since many practical well tests are performed using only a single monitoring well in which drawdown is recorded as a function of time, some workers prefer to plot the well function versus 1/u as shown in the bottom panel in Figure 14.3. In this format, at a fixed radius, time increases to the right, and the plot more closely resembles the actual drawdown in a well with time.

Drawdown in Space and Time

An example of the drawdown of the hydraulic head associated with a continuously discharging well is presented in Figure 14.4 where we have used the same aquifer parameters as for the impulse response discussed above, and for the steady-state flow case discussed in Chapter 10. In the top panel we show the behavior of h as a function of distance at various times (t = 10, 100, 1000 min). In the bottom panel, we see the behavior of h as a function of time at several fixed distances from the discharging well (r = 25, 50 and 100 ft; or 7.6 m, 15.2 m and 30.5 m, respectively).

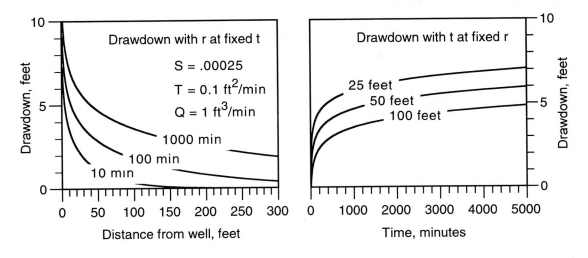

Figure 14.4 Example of the drawdown of the hydraulic head for constant discharge (Q = 1.0 ft^3/min) beginning at t = 0, for the same aquifer parameters used for Figure 14.2. **Left panel:** Behavior of h with r at fixed t = 10, 100, 1000 min. **Right panel:** Behavior of h with t at fixed r = 25, 50 and 100 ft.

Asymptotic Form for Large Times or Small Distances

Note that for small u (say u ≤ 0.05) — corresponding to small r or large t, respectively — one can neglect the summation term in (14.47) and simply write

$$E_1(u) \approx -0.5772 - \ln u \qquad (14.48)$$

As mentioned above, in some well tests it is often more convenient to plot $E_1(u)$ as a function of 1/u, so that as time increases the parameter 1/u increases to the right. Using the identity that

$$\ln(u) = -\ln(1/u) \qquad (14.49)$$

and combining the two terms in (14.48) into the argument of a single logarithmic function, we obtain

$$E_1(u) \approx \ln[2.246(Tt)/(Sr^2)] \qquad (14.50)$$

where we have used $1/u = [(4Tt)/(Sr^2)]$, and that

$$\text{antiln}(-0.5772) = 0.5615 \qquad (14.51)$$

which in words states that the anti-logarithm of – 0.5772 to the base e is 0.5615.

In addition, for plotting and analysis purposes, it is usual to use logarithms to the base 10, so that by introducing the identity

$$\log C = \ln C \cdot \log e \qquad (14.52)$$

(14.50) becomes

$$E_1(u) \approx 2.302 \log[2.246\,(Tt)/(Sr^2)] \tag{14.53}$$

Results in Terms of the Perturbation of the Hydraulic Head

Upon substituting (14.53) for the integral on the right hand side of (14.46), we obtain the following approximate expression for the perturbation in hydraulic head in the limit of small u (corresponding to small r or large t)

$$\delta h(r,t) \approx -(2.3Q/4\pi T) \log[2.25\,(Tt)/(Sr^2)] \tag{14.54}$$

Results in Terms of Drawdown

Some authors express the results of well tests in terms of a drawdown parameter defined by

$$\text{Drawdown} = h_o - h(r,t) \tag{14.55}$$

which is the negative of our $\delta h(r,t)$. In this case, (14.46) would lead to the general expression

$$h_o - h = (Q/4\pi T)\,E_1(u). \tag{14.56}$$

The asymptotic form of (14.56) for large t or small r would accordingly be

$$h_o - h \approx (2.3Q/4\pi T) \log[2.25\,(Tt)/(Sr^2)] \tag{14.57}$$

The drawdown relation (14.57) can be expanded to the form

$$h_o - h \approx (2.3Q/4\pi T) \log[2.25\,(T/S)]$$
$$+ (2.3Q/4\pi T) \log(t) - 2\,(2.3Q/4\pi T) \log(r) \tag{14.58}$$

which, as we will see in the following section, is the equation of a plane in the three dimensional space of drawdown, time and distance.

Case A: Drawdown as a function of time at fixed radius. For drawdown as a function of time in a well at a fixed distance, (14.58) is the equation of a straight line

$$y = mx + b \tag{14.59}$$

in semi-log coordinates. This may be seen by setting

$$y = h_o - h \tag{14.60a}$$
$$b = (2.3Q/4\pi T) \log[2.25\,T/(Sr^2)] \tag{14.60b}$$
$$m = 2.3Q/(4\pi T) \tag{14.60c}$$
$$x = \log(t) \tag{14.60d}$$

The asymptotic form of (14.61) is plotted against a "complete" response of $(h_o - h)$ vs $\log(t)$ at fixed radius in the top panel of Figure 14.5.

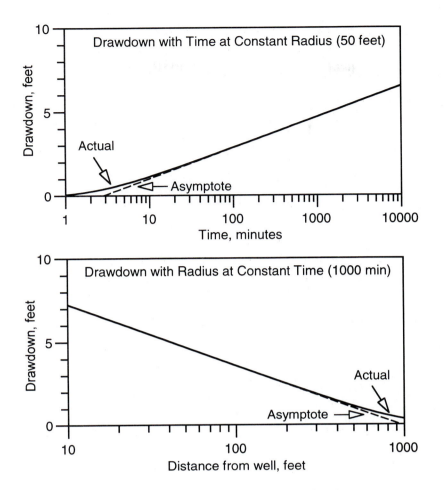

Figure 14.5 *Top panel:* The asymptotic form of (14.61) plotted against a "complete" response of $(h_o - h)$ vs log (t). *Bottom panel:* The asymptotic form of (14.62) plotted against a "complete" response of $(h_o - h)$ vs log (r).

Case B: Drawdown as a function of radius at fixed time. For drawdown as a function of radius at a fixed time, (14.58) is the equation of a straight line (14.59) providing we set

$$y = h_o - h \qquad\qquad (14.61a)$$
$$b = (2.3Q/4\pi T) \log [2.25\,T\,t/S] \qquad\qquad (14.61b)$$
$$m = -(4.6Q/4\pi T) \qquad\qquad (14.61c)$$
$$x = \log (r) \qquad\qquad (14.61d)$$

The asymptotic form of (14.61) is plotted against a "complete" response of $(h - h_o)$ vs log (r) at a fixed time in the bottom panel of Figure 14.5.

Jacob's Straight Line Method

A simple, but elegant analysis to determine aquifer properties has been developed using the asymptotic forms for large t or small r represented by (14.54) or (14.57) above. The technique, named for its originator, is known for reasons that are apparent from the last section as "Jacob's

- 213 -

straight line method". (Some workers refer to the procedure as the "Cooper-Jacob straight line method.)"

Straight line curve fits. For either Case A or Case B, above, Jacob's method consists of "fitting" a straight line to field data, at small r or large t, plotted as $(h_o - h)$ versus the dependent variable log [t] or log [r], using semi-log paper. This allows one to determine T and S from the slope of the line and the intercept $y = 0$.

"Complete" Curve Matching

To reiterate some of what has been discussed above, historically, hydrogeologists have represented the exponential integral as the well function W(u), such that

$$W(u) = E_1(u) \tag{14.62}$$

and the drawdown

$$h_o - h = (Q/4\pi T)\, W(u) \tag{14.63}$$

where the Theis parameter, u, is

$$u = r^2 S / 4 t T \tag{14.44}$$

For fixed-time variable-distance well tests using multiple monitoring wells, field data are plotted as $(h_o - h)$ versus r^2/t, and compared to an invariant curve of W(u) versus u. When a "match" (or curve fit) is obtained, values of T and S are calculated from the appropriate common match points — that is by determining the set of values for W(u) and u corresponding to the observed values for $(h_o - h)$ and r^2/t. Alternatively, for fixed distance variable time well tests, such as measuring drawdown with time in a single well, field data might be plotted as $(h_o - h)$ versus t/r^2 and compared to an invariant curve of W(u) versus $1/u$.

While this was done by many generations of hydrogeologists using transparent overlays, nowadays such data fits are often achieved by computer algorithms.

Chapter 15. Selected Topics in Transient Flow

REVIEW OF TRANSIENT FLOW RELATION IN THREE DIMENSIONS

Fundamental Relations

The conservation of flux through an elemental volume of a confined aquifer leads to the divergence condition:

$$\nabla \cdot \mathbf{q} = -S_s \, \partial h / \partial t \qquad (15.1)$$

where $\mathbf{q}$ is the specific discharge vector, h is the hydraulic head and S_s is the specific storage coefficient; the latter being the water released from storage in a unit volume due to a unit decrease in the hydraulic head.

Darcy's law for an isotropic medium is

$$\mathbf{q} = -K \, \nabla h \qquad (15.2)$$

K being the hydraulic conductivity.

Diffusion Equation for Hydraulic Head

Upon substituting the right side of (15.2) for $\mathbf{q}$ on the left side of (15.1), we obtain the following form for the diffusion equation in terms of the hydraulic head:

$$\left[\frac{\partial}{\partial x} \left(K \frac{\partial h}{\partial x} \right) + \frac{\partial}{\partial y} \left(K \frac{\partial h}{\partial y} \right) + \frac{\partial}{\partial z} \left(K \frac{\partial h}{\partial z} \right) \right] = S_s \frac{\partial h}{\partial t} \qquad (15.3)$$

which is the fundamental relation for transient flow in three dimensions, and has the form of a diffusion equation

Assuming K is homogeneous, the diffusion equation for transient flow in a confined aquifer becomes

$$\frac{\partial^2 h}{\partial x^2} + \frac{\partial^2 h}{\partial y^2} + \frac{\partial^2 h}{\partial z^2} = \frac{S_s}{K} \frac{\partial h}{\partial t} \qquad (15.4)$$

TWO DIMENSIONAL TRANSIENT DISCHARGE POTENTIALS FOR CONFINED AND UNCONFINED FLOW

Darcy's Law in the Discharge Potential

We consider here confined flow in the horizontal plane as well as unconfined flow under the Dupuit assumptions discussed earlier in the text. For either of these two cases, Darcy's law has the following components

$$q_x = -K \, \partial h / \partial x \qquad (15.5)$$

$$q_y = -K \, \partial h / \partial y \qquad (15.6)$$

We assume that flow will be distributed uniformly over the thickness b of a confined aquifer, or over the thickness of the saturated section h for an unconfined aquifer. In either case flow is confined to an effective thickness that we will denote by b_{eff}. Multiplying both sides of (15.5) and (15.6) by the effective thickness of the aquifer b_{eff}, as in our earlier discussion, we can define the horizontal discharge vector

$$\mathbf{Q} = (Q_x, Q_y) \qquad (15.7)$$

where

$$Q_x = b_{eff}\, q_x \qquad (15.8)$$

$$Q_y = b_{eff}\, q_y \qquad (15.9)$$

Upon substituting the Darcy's law expressions embodied in (15.5) and (15.6) for q_x and q_y on the right hand sides of (15.8) and (15.9), respectively, the latter become

$$Q_x = b_{eff}\, q_x = -\, b_{eff}\, K\, \partial h/\partial x \qquad (15.10)$$

$$Q_y = b_{eff}\, q_y = -\, b_{eff}\, K\, \partial h/\partial y \qquad (15.11)$$

For *confined* flow we define the discharge potential Φ_c by

$$\Phi_c = b_{eff}\, K\, h + C_c \qquad (15.12)$$

or

$$\Phi_c = b\, K\, h + C_c \qquad (15.13)$$

For *unconfined* flow we define the discharge potential Φ_u by

$$\Phi_u = (b_{eff}\, K\, h)/2 + C_u \qquad (15.14)$$

or

$$\Phi_u = (K h^2)/2 + C_u \qquad (15.15)$$

For either types of flow, we see that Darcy's law, (15.10) and (15.11), can be rewritten in terms of the discharge potential

$$Q_x = -\, \partial\Phi/\partial x \qquad (15.16)$$

$$Q_y = -\, \partial\Phi/\partial y \qquad (15.17)$$

Conservation Condition on the Discharge Potential

The conservation condition on the horizontal discharge vector $\mathbf{Q}$ is given in general by

$$\partial Q_x/\partial x + \partial Q_y/\partial y = b_{eff}\, w \qquad (15.18)$$

where b_{eff} is b for a confined aquifer and h for an unconfined aquifer. The product $b_{eff}\, w$ represents the total amount of water produced from sources in the aquifer in a vertical column of unit cross sectional area in the horizontal plane.

Substituting (15.16) and (15.17), respectively, for Q_x and Q_y in (15.18), we have the following diffusion equation for the discharge potential for either confined or unconfined flow

$$\partial^2 \Phi / \partial x^2 + \partial^2 \Phi / \partial y^2 = - b_{eff} w \qquad (15.19)$$

where $b_{eff} w$ represents the water produced from local sources over the thickness of the aquifer.

Fluid Production Associated with Transient Changes in h

We now consider the case where the primary source of water in the aquifer is the release of fluid from storage. Moreover, we assume, as discussed in Chapter 11, that the release of stored water is directly related to a change in the hydraulic head.

Confined aquifer: Storage properties. As discussed in Chapter 11, for a *confined* aquifer, the release of water associated with a change of hydraulic head is contributed by a combination of the elastic *response* of the fluid and aquifer matrix. Quantitatively, these effects are embodied in the specific storage given by

$$S_S = [\alpha + n\beta] \rho g, \qquad (15.20)$$

where α is the matrix compressibility, n is the porosity, β is the fluid compressibility, ρ is the density of water and g is the force of gravity per unit mass. Thus, for confined flow, the source term on the right hand side of (15.19) becomes

$$b_{eff} w = - S_s b \, \partial h / \partial t = - S \, \partial h / \partial t \qquad (15.21)$$

where S_s is the specific storage and $S = S_s b$ is the storativity of the confined aquifer. Recalling the definition of the discharge potential for confined flow given by (15.13), and dropping the constant C_c as being nonessential, (15.21) can be rewritten as

$$b_{eff} w = - S \, \partial h / \partial t = - (S/T) \, \partial \Phi_c / \partial t \qquad (15.22)$$

Unconfined aquifer: Storage properties. We also discussed in Chapter 11 that the storativity of an *unconfined* aquifer is given by

$$S = S_y + h S_s \qquad (15.23)$$

where, in addition to the elastic term S_s defined by (15.20), we have the specific yield S_y of the aquifer due to dewatering of the saturated section from gravity drainage as the water table, h, is drawn down. The specific yield is given as the amount of water released per unit area for a unit decrease in the height of the water table.

Recall from Chapter 11 that typical values of the specific yield (S_y) for unconfined aquifers are 0.01 to 0.3, whereas the specific storage coefficient (S_s) is on the order of .005 to .00005. Thus, even though the specific storage coefficient is multiplied by the thickness h of the saturated section in (15.23), this term is usually — though not invariably — much smaller than the specific yield S_y, and is often neglected in some analyses.

In general, however, both processes contribute to the response of an unconfined aquifer to a transient change in hydraulic head, often at different stages of a transient event.

Confined Flow

Diffusion equation for Φ_c. Upon substituting the storage release term (15.22) for the source term on the right hand side of (15.19), we obtain the following fundamental form for transient flow in a confined aquifer

$$\partial^2\Phi_c/\partial x^2 + \partial^2\Phi_c/\partial y^2 = (S/T)\,\partial\Phi_c/\partial t \qquad (15.24)$$

Diffusion equation for Φ_c in cylindrical coordinates. In two-dimensional cylindrical coordinates (also known as "circular" coordinates, since independence of z is assumed), (15.24) can be written as

$$\frac{\partial^2\Phi_c}{\partial r^2} + \frac{1}{r}\frac{\partial\Phi_c}{\partial r} = \frac{S}{T}\frac{\partial\Phi_c}{\partial t} \qquad (15.25)$$

where Φ_c is defined by (15.13). This is precisely the form for the diffusion equation for the hydraulic head

$$\frac{\partial^2 h}{\partial r^2} + \frac{1}{r}\frac{\partial h}{\partial r} = \frac{S}{T}\frac{\partial h}{\partial t} \qquad (14.4)$$

that we developed in Chapter 14, and which has the solution

$$\delta h(r,t) = -(Q/4\pi T)\,E_1(u) \qquad (14.46)$$

From the definition of $\Phi_c(r,t)$ by (15.13), following Strack (1989), multiplying both sides of the latter relation by the transmissivity T, and noting from the definition in (13.13) that $\delta\Phi_c(r,t) = T\,\delta h(r,t)$, we obtain

$$\delta\Phi_c(r,t) = -(Q/4\pi)\,E_1(u) \qquad (15.26)$$

which is the solution for the discharge potential for confined flow to a well that commenced pumping at a constant rate Q at $t = 0$.

Unconfined Flow

Upon substituting the storativity for an unconfined aquifer given by (15.23) for S in the last term of (15.21), the latter can replace the right hand side of (15.19), resulting in

$$\partial^2\Phi_u/\partial x^2 + \partial^2\Phi_u/\partial y^2 = S\,\partial h/\partial t \qquad (15.27)$$

which is a *nonlinear* differential equation because of the mix of terms in h to the first power on the right hand side and the term involving h to the second power (implied in Φ_u) on the left hand side. In order to "linearize" (15.27), so that we can employ the powerful techniques of linear differential equations to solve it, we need to manipulate the relation to a form that involves only terms to the first power in Φ_u.

Linearizing the diffusion equation for unconfined flow. Equation (15.27) can be linearized by first noting that the storativity S, for unconfined flow, is a function of h according to (15.23). This implies that the source term on the right side of (15.27) becomes

$$S\frac{\partial h}{\partial t} = \left(S_y + h S_s \right)\frac{\partial h}{\partial t}$$

(15.28)

which can be factored to the form

$$\left(S_y + h S_s \right)\frac{\partial h}{\partial t} = S_y\frac{\partial h}{\partial t} + S_s\frac{1}{2}\frac{\partial h^2}{\partial t}$$

(15.29)

Recalling the definition of the discharge potential for unconfined flow given by (15.15), we see that the last term in h^2 on the right of (15.29) can be rewritten in terms of Φ_u, so that the entire right hand side of this expression becomes

$$= S_y\frac{\partial h}{\partial t} + \frac{S_s}{K}\frac{\partial \Phi_u}{\partial t}$$

(15.30)

The second term on the right is now "linearized" without approximation. As for the first term on the right, multiply the numerator and denominator by the local transmissivity of the saturated section

$$T = h \cdot K$$

(15.31)

we obtain

$$S_y\frac{\partial h}{\partial t} = S_y\frac{\partial h}{\partial t}\frac{Kh}{Kh}$$

(15.32)

which can be rewritten in terms of Φ_u as

$$S_y\frac{\partial h}{\partial t} = \frac{S_y}{Kh}\frac{\partial \Phi_u}{\partial t}$$

(15.33)

Now h, in the denominator of the coefficient on the right hand side, can be expressed as the sum of the initial static head h_o and a transient disturbance δh, such that

$$h = h_o + \delta h = h_o(1 + \varepsilon)$$

(15.34)

where we have used ε in the last term to represent the perturbation

$$\varepsilon = \delta h/h_o$$

(15.35)

According to the binomial theorem, for small ε, the quotient can be approximated by

$$\frac{1}{(1 + \varepsilon)} \approx (1 - \varepsilon)$$

(15.36)

so that (15.33) can be written as

$$S_y \frac{\partial h}{\partial t} = \frac{S_y}{Kh_o} \frac{\partial \Phi_u}{\partial t} (1 - \varepsilon) \qquad (15.37)$$

Upon neglecting terms of order $\delta h / h_o$, the latter expression becomes

$$S_y \frac{\partial h}{\partial t} \approx \frac{S_y}{T_o} \frac{\partial \Phi_u}{\partial t} \qquad (15.38)$$

where $T_o = Kh_o$. The entire source term expression (15.29) thus becomes

$$\left(S_y + h S_s \right) \frac{\partial h}{\partial t} = \frac{S_y}{T_o} \frac{\partial \Phi_u}{\partial t} + \frac{S_s}{K} \frac{\partial \Phi_u}{\partial t} \qquad (15.39)$$

which, upon multiplying the numerator and denominator of the second term on the right by h_o, can be factored to the form

$$= \left(\frac{S_y}{T_o} + \frac{h_o S_s}{h_o K} \right) \frac{\partial \Phi_u}{\partial t} = \left(\frac{S_y + h_o S_s}{T_o} \right) \frac{\partial \Phi_u}{\partial t} \qquad (15.40)$$

We now define what one might term the ratio of the unconfined storativity S_u to the unconfined transmissivity T_u by the operation

$$\frac{S_u}{T_u} = \frac{S_y + h_o S_s}{T_o} \qquad (15.41)$$

Note that this ratio is an operational definition, and the two parameters — the storativity and the transmissivity — are not decoupled as separate physical entities as they were for the confined aquifer case. With this definition, the right hand term in (15.27) becomes

$$S \frac{\partial h}{\partial t} = \frac{S_u}{T_u} \frac{\partial \Phi_u}{\partial t} \qquad (15.42)$$

so that (15.27) is "linearized" as long as the relative perturbation $\delta h / h_o$ is small, and the definition (15.41) is presumed. We will assume henceforth that $(S_u/T_u) \approx (S/T)$. In this case, (15.27) becomes

$$\partial^2 \Phi_u / \partial x^2 + \partial^2 \Phi_u / \partial y^2 = (S/T) \partial \Phi_u / \partial t \qquad (15.43)$$

which is identical in form to the relation for confined flow (15.24); except, of course, for the intrinsic differences in the definitions of the respective discharge potentials. This similarity permits very powerful procedures for solving the unconfined flow problem as an analog to the confined flow problem (Strack, 1989).

Diffusion equation for Φ_u in cylindrical coordinates. In two-dimensional cylindrical coordinates (15.43) becomes

$$\frac{\partial^2 \Phi_u}{\partial r^2} + \frac{1}{r}\frac{\partial \Phi_u}{\partial r} = \frac{S}{T}\frac{\partial \Phi_u}{\partial t} \qquad (15.44)$$

which as we showed with reference to (15.25) is precisely the form of the diffusion equation for the discharge potential for confined flow.

Response to discharge from a well pumping an unconfined aquifer. For the case of the transient drawdown of an unconfined aquifer to constant discharge Q from a well commencing at $t = 0$, (15.44) has a solution in terms of the exponential integral $E_1(u)$. Thus a solution to the *unconfined* flow problem in terms of the discharge potential defined by (15.15), analogous to the solution for *confined* flow given by (15.26), is given by

$$\delta\Phi_u(r,t) = - (Q/4\pi)\, E_1(u) \qquad (15.45)$$

Note that the constant C_u cancels when we calculate the difference term $\delta\Phi_u(r,t)$, so that, in terms of the hydraulic head, (15.45) becomes

$$h^2 - h_o^2 = -(Q/2\pi K)\, E_1(u) \qquad (15.46)$$

In addition to the "complete" form given by (15.45), we have the approximate expression for large t or small r given by

$$\delta\Phi(r,t) \approx - (2.3\,Q/4\pi)\, \log\,[2.25\,(Tt)/(Sr^2)] \qquad (15.47)$$

from analogy with the discussion in Chapter 14 (see expression (14.54)). In terms of hydraulic head, becomes

$$h^2 - h_o^2 \approx - (2.3\,Q/2\pi K)\, \log\,[2.25\,(Tt)/(Sr^2)] \qquad (15.48)$$

Example of a Well Test on an Unconfined or Water Table Aquifer

Stages of drawdown. According to Boulton (1963), and as reviewed by Freeze and Cherry (1979) and Fetter (1994), the flow of water to a well discharging an unconfined aquifer proceeds in three stages:

- In early times, water is released from elastic storage through aquifer compaction and water decompression similar to confined aquifers.

- During intermediate times, fluid is actually drained from the aquifer as the water table is lowered, but some water is retained above the water table, and its release is delayed. [This is known as "delayed yield" and, during this phase, a substantial source of water may remain above the water table.]

- In later times, gravity drainage (dewatering) of the aquifer is in equilibrium with the fall of the water table.

Application to actual data. The difference between the early time response and the late time response of an unconfined aquifer to a discharging well may be seen in Figure 15.1 which compares well test data from the Hudson River Valley in Saratoga, NY. The actual data are selected from a suite of measurements reported by Heath and Trainer (1992), and the theoretical responses were calculated from the above considerations.

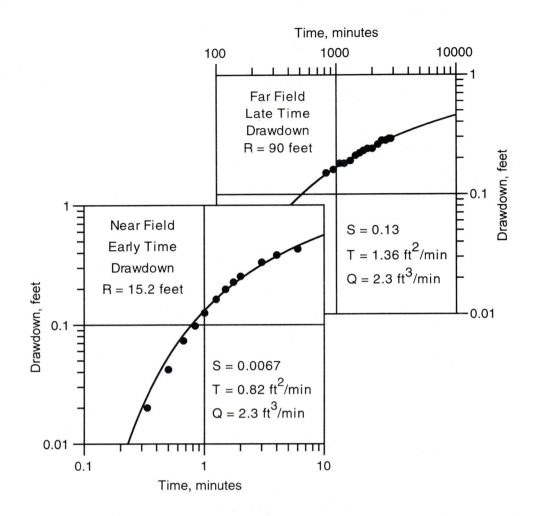

Figure 15.1 Comparing observed drawdown data from an unconfined aquifer to theoretical responses at early time and late time. The early time storativity (.0067) appears to be dominated by the release of water from elastic storage. The late time storativity (0.13) appears to be dominated by the specific yield and actual dewatering of the aquifer. (Data are from the Hudson River Valley in Saratoga, NY; Heath and Trainer, 1992.)

The lower left panel represents data obtained from a monitoring well at times on the order of one minute after pumping began in the discharge well 15.2 ft (4.6 m) away. We refer to these observations as "near-field, early-time" data. The upper right panel represents data obtained from a monitoring well at times on the order of 1000 minutes after pumping began in the discharge well 90 ft (27.4 m) away. We refer to these observations as "far-field, late-time" data. A discharge rate of 2.3 ft^3 min^{-1} corresponds to a value of 94 m^3 day^{-1} in SI units.

To interpret these results, a Jacob's straight-line fit to each data set produced initial estimates for S and T for the respective wells. These were then adjusted to produce the response curves shown in Figure 15.1 based on (15.46). The transmissivity increases somewhat during the pumping sequence from an estimate of 0.82 ft^2 min^{-1} (110 m^2 day^{-1}) in early time, to a value of 1.36 ft^2 min^{-1} (182 m^2 day^{-1}) at late time. However, the most dramatic change occurs in the *storativity* which increases by a factor of almost 20, from a value of 0.0067 in the near field at early time to approximately 0.13 in the far field at late time. The former value of storativity is consistent with a mechanism by which water is released through decompression of the matrix and the interstitial fluid. The latter is consistent with water released through specific yield and the actual dewatering of the aquifer as the water table is lowered through drawdown. This is the effect of "gravity drainage".

The difference between the response of these two mechanisms in space and time can be appreciated from inspecting Figure 15.2, where we see the "complete" response curves represented by the two extremes in Figure 15.1.

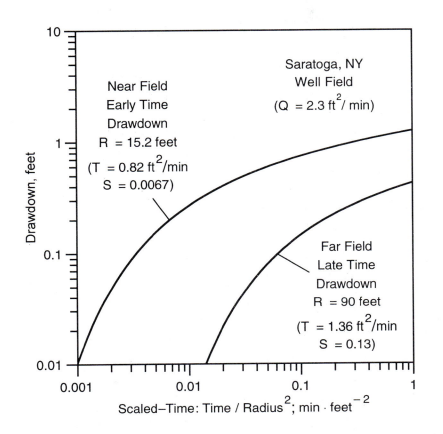

Figure 15.2 Comparing the early time and late time responses of the unconfined aquifer discussed in the text by plotting drawdown as a function of the *scaled-time*, calculated by dividing the actual time of observation by the square of the distance between the discharging well and the respective observation well.

In order to compare the data from the two wells, we use the "scaled-time" t/r^2. That is to say, the time for each observation in minutes has been normalized, or scaled, by the square of the distance in feet from the monitoring well to the discharging well. If, for this particular well test, the storativity and transmissivity were invariant with time and distance then the far field, late time data would have overlaid the near field, early time data when both data sets were plotted in *scaled-time*. The fact that the curves *do not* overlay, is inferred to be due to the significantly longer time delay intrinsically associated with the large storativity available from gravity drainage; there is a substantial offset of this response toward later times.

While the above piece-wise approach, or variations on it, are satisfactory for many applications, the reader is referred to more elegant, comprehensive procedures following from the initial theoretical work of Boulton (1963) and Neuman (1975), and summarized by Freeze and Cherry (1979), Walton (1987), Watson and Burnett (1993) and Fetter (1994).

REFERENCES AND RECOMMENDED READING

Abramowitz, Milton, and Irene A. Stegun, *Handbook of Mathematical Functions*, 1046 pp., Dover Publications, New York, 1965.

Anderson, Mary P. and William W. Woessner, *Applied Groundwater Modeling*, 381 pp., Academic Press, New York, 1992.

Bear, Jacob, *Dynamics of Fluids in Porous Media*, 764 pp., Dover Publications, New York, 1988.

Boulton, N,S., Analysis of data from non-equilibrium pumping tests allowing for delayed yield from storage, *Proceed. Inst. Civil Engrs.* (London), *26*, p. 469-482, 1963.

Bras, R.L., *Hydrology*, 643 pp., Addison Wesley Publishing Co., Reading, MA, 1990.

Carslaw, H.S., and J.C. Jaeger, *Conduction of Heat in Solids*, Oxford University Press, 510 pp., 1959.

Chow, V.T., D.R. Maidment, and L.W. Mays, *Applied Hydrology*, 572 pp., McGraw-Hill, Inc., New York, 1988.

Davis, Stanley N., and Roger J. M. DeWiest, *Hydrogeology*, 463 pp., John Wiley and Sons, Inc., New York, 1966.

Dingman, S.L., *Physical Hydrology*, 575 pp., Macmillan Publishing Co., New York, 1994.

Domenico, Patrick A., and Franklin W. Schwartz, *Physical and Chemical Hydrogeology*, Second Edition, 506 pp., John Wiley and Sons, Inc., New York, 1998.

Driscoll, Fletcher G., *Groundwater and Wells*, Second Edition, 1108 pp., Johnson Filtration Systems, Inc., St. Paul, MN, 1986.

Fetter, C.W., *Applied Hydrogeology*, 691 pp., Third Edition, Macmillan Publishing Co., New York, 1994.

Fetter, C.W., *Contaminant Hydrogeology*, 458 pp., Macmillan Publishing Co., New York, 1993.

Freeze, R. Allan, and John A. Cherry, *Groundwater*, 604 pp., Prentice-Hall, Englewood Cliffs, NJ, 1979.

Gradshteyn, I.S. and I.M. Ryzhik, *Tables of Integrals, Series and Products*, 1086 pp., Academic Press, New York, 1965.

Heath, Ralph C., *Ground-Water Regions of the United States*, United States Geological Survey Water-Supply Paper 2242, 78 pp., United States Government Printing Office, 1984.

Heath, Ralph C., *Basic Ground-Water Hydrology*, United States Geological Survey Water-Supply Paper 2220, United States Government Printing Office, 1984.

Heath, Ralph C., and Frank W. Trainer, *Introduction to Ground Water Hydrology*, 285 pp., National Ground Water Association, Dublin, OH, 1992.

Hubbert, M. King, The theory of ground-water motion, *J. Geology*, *48*, 8, Part I, p. 785-944, 1940.

Kaplan, W., *Advanced Calculus*, 679 pp., Addision-Wesley Publishing Co., Reading, MA, 1952.

Kazmann, R.G., *Modern Hydrology*, Third Edition, 427 pp., National Water Well Association (now National Ground Water Association), Dublin, OH, 1988.

Landau, L.D., and E.M. Lifshitz, *Fluid Mechanics*, Addison-Wesley Publishing Co., Reading, MA, 1959.

Lebedev, N.N., *Special Functions and Their Applications*, 308 pp., Dover Publications, New York, 1972.

Margenau, H., and G.M. Murphy, *The Mathematics of Physics and Chemistry*, Second Edition, 604 pp., Robert E. Krieger Publishing Co., Huntington, NY, 1976.

de Marsily, Ghislain, *Quantitative Hydrogeology*, 440 pp., Academic Press, San Diego, 1986.

Mathews, J., and R.L. Walker, *Mathematical Methods of Physics*, Second Edition, 501 pp., W.A. Benjamin, Inc., New York, 1970.

Morse, Philip, and Herman Feshbach, *Methods of Theoretical Physics*, Volumes I and II, 1978 pp., McGraw-Hill Book Co., New York, 1953.

Neuman, S.P., Theory of flow in unconfined aquifers considering delayed response to the water table, *Water Resources Res.*, 8, p. 1031-1045, 1972.

Neuman, S.P. and P.A. Witherspoon, Applicability of current theories of flow in leaky aquifers, *Water Resources Res.*, 5, p. 817-829, 1969.

Roscoe Moss Company, *Handbook of Groundwater Development*, 493 pp., John Wiley and Sons, New York, 1990.

Sen, Zekai, *Applied Hydrology for Scientists and Engineers*, 444 pp., Lewis Publishers, Boca Raton, FL, 1995.

Sokolnikoff, I.S., and Redheffer, *Mathematics of Physics and Modern Engineering*, 812 pp., McGraw-Hill Book Co., New York, 1958.

Strack, O.D.L., *Groundwater Mechanics*, 732 pp., Prentice Hall, Englewood Cliffs, NJ, 1989.

Theis, C.V., The relation between the lowering of the piezometric surface and the rate and duration of discharge of a well using ground-water storage, *Trans. Am. Geophys. Union*, 16, p. 519-524, 1935.

Todd, Keith David, *Groundwater Hydrology*, 535 pp., John Wiley and Sons, New York, 1980.

Toth, J.A., A theoretical analysis of ground-water flow in small drainage basins, *J. Geophys. Res.*, 68, p. 4795-4812, 1963.

Walton, W.C., *Practical Aspects of Groundwater Modeling*, Third Edition, 587 pp., National Water Well Association (now National Ground Water Association), Dublin, OH, 1988.

Walton, W.C., *Analytical Groundwater Modeling: Flow and Contaminant Migration*, 173 pp., Lewis Publishers, Boca Raton, FL, 1989.

Walton, W.C., *Numerical Groundwater Modeling: Flow and Contaminant Migration*, 272 pp., Lewis Publishers, Boca Raton, FL, 1989.

Walton, W.C., *Principles of Groundwater Engineering*, 546 pp., Lewis Publishers, Boca Raton, FL, 1991.

Wang, Herbert F., and Mary P. Anderson, *Introduction to Groundwater Modeling: Finite Difference and Finite Element Methods*, 237 pp., W. H. Freeman and Company, San Francisco, 1982.

Watson, I., and A.D. Burnett, *Hydrology - An Environmental Approach*, 702 pp., Buchanan Books Cambridge, Ft. Lauderdale, FL, 1993.

INDEX

Depth of penetration, 166
Determining h(x,y) from V(x,y), 73
Determining q from V(x,y), 72
Dewatering, as storage loss, 158
Diagonalized, conductivity tensor, 22
Differential circulation, definition, 55
   for rigid rotation, 57
Differential pressure, and effective
   stress, 149
Diffusion equation for transient flow:
   discussion, 154, 200
   discharge potential, confined or
     unconfined flow, 216
   radial flow discharge potential, 218,
     220, 221
Diffusion, examples, 185, 186
Dimensionality of practical
   hydrogeological situations, 23
Dirac delta function, 172, 173, 206
Direction cosines, 32, 33, 62
Direction of the hydraulic gradient, 33
Directional finite difference, 32
Directional derivatives, 26, 32
Discharge potential, for radial flow,
   confined flow, 128
   boundary condition on, 100, 101
   unconfined flow, 99
   radial flow, solution, 129
   1-D steady-state flow, unconfined
     aquifer, examples, 101, 105
   transient confined flow to a well, 218
   2-D horizontal confined flow, 124
   unconfined radial flow, 135
   unconfined flow, 219
Discharge, system, simple spatial
   for vertical 2-D system, 109
   from well, convention for sign, 119
   radial flow, 204
Discontinuity in q tangential, 87
Discrete impulses, summation of, 184
Discrete superposition, 184
Divergence-free condition, and flow
   functions, 60
   on specific discharge q, 72
   and source free condition, 54
Divergence condition, derivation with
   sources, in 3-D, 53
   on total discharge vector, 2-D
     unconfined flow, 99
   time dependent source terms, 144
   unconfined radial flow, 135
Divergence, for 2-D flow functions, 60
   for 1-D unconfined flow, 95
   of flux, derivation, 49
   of flux, 3-D, 54, 57, 58
   of radial confined flow to a well, 127
   of radial unconfined flow, 135
   for total discharge vector, and 2-D
     horizontal flow, 124
Drawdown due to discharging well:
   steady-state relation, 131

in space and time, transient well
   discharge, 210
of hydraulic head, steady-state, 131
asymptotic form for late times or
   small distances, 211
in the discharging well, for radial
   unconfined flow, 138
of discharge potential, for radial
   unconfined flow, 137
of h, for radial unconfined flow, 138
Δh as function of t at fixed r, 212
Δh as function of r at fixed t, 213
Dupuit assumptions, 1-D steady-state
   flow, unconfined aquifer, 102
   condition to be valid, 94
   discussion, 92, 93
   unconfined flow to well, 133
   for transient unconfined flow, 158
Dupuit flow, in unconfined aquifers,
   discussion, 92, 94
   condition on h-lines, 95
Dynamic estimate of T, 132
Dynamic viscosity, fluid property, 9, 10
Early time, transition to steady state
   flow to discharging well, 120
Effective pressure, same as effective
   stress, 149
   relation to change in h, 150
Effective stress, in terms of h, 149
   and differential pressure, 149
   change, related to change in h, 151
   change, sign relative to Δh, 150
   discussion, 148
Eigenfunction, eigenvalue relation, and
   eigenvalue, for Laplace's eqn., 115
Elastic behavior of the subsurface, 145
Elastic properties of aquifer,
   importance to storage, 150
Elastic storage, 201, 204
Electric current density, analogy with
   fluid flux, 9, 69
Electric displacement field, analogy
   with fluid flux, 36
Electric field intensity, analogy, 9
Electric voltage, analogy with h, 69
Electrical conductivity, analogy, 9
Electricity and magnetism, and stream
   functions, 59
Elementary functions, 179
Elementary impulse response, 188
Elementary solutions of transient flow
   equation, 162
Elevation head, definition, 17
Energy, and fluid potential, 18
   associated with pressure, 18
   gravitational, 18
   kinetic, 18
   total mechanical, 18
Equilibrium flow, for well tests,
   determined using two wells, 122
   stabilized, 121, 122, 131

Equipotential line, h-line, 69
Equipotentials, in electricity, 69
   of hydraulic head, 69
Error function, 189, 190, 191, 210
Evaporation, 3, 4
Exfiltration, water loss, 4
Exp(u), exponential of u to base e, 170
Exponential attenuation, 165
   of fluid circulation, 118
   of potential function with depth, 118
   of specific discharge with depth, 117
Exponential integral, 209, 210, 214,
   221, 164
Exponential term, as solution of
   Laplace's equation, 115
   decrease/increase with distance, 116
Expulsion of water, from an elastic
   matrix, 151
External stress, role in compaction, 146
Extrapolation of a function, 26, 27
   problems with, 28
   need for higher order terms, 28
Extrapolation, of flux, 37
Fall line, opposite of elevation
   gradient, 34
Fetter, 15, 44, 77, 94, 100, 132, 160,
   221, 223
Filtration velocity, q, 8
Finite length confined aquifer, 178
Flow-line, orthogonal to h-line, 60
   indicates local flow direction, 34
Flow-net, grid of flow-lines and h-
   lines, 60
   analytical basis for, 60
   assumptions for constructing, 60
Flow equation, 3-D with sources, 58
Flow function, V(x,y), 59, 61
   as scalar potential function, 66
   stream function, synonymous, 59
Flow in distributed media, 20
Flow in absence of local sources, 100
Flow line, 67–70
   as V(x,y) = constant, 68
Flow net, analytical example of, 71
   discussion, 59
   quantitative interpretation, 71
   rules for classical construction, 70
Flow relation, 1-D unconfined flow, 95
   for discharge potential, 99
   2-D unconfined flow, 98
Flow with local sources, 1-D steady-
   state flow, unconfined aquifer, 104
Fluid circulation, shallow vs. deep, 118
Fluid potential, and Darcy's law, 19
   and total mechanical energy, 18
Fluid pressure, and energy, 18
   at watertable, 13
Fluid property, dynamic viscosity, 9
   weight density, 9
Flux q, definition, 1, 8
Flux through a closed contour, 64